SCIENTISTS,
BUSINESS,
AND THE STATE,
1890–1960

THE
LUTHER
HARTWELL
HODGES
SERIES

ON
BUSINESS,
SOCIETY,
AND THE
STATE

WILLIAM H. BECKER, EDITOR

# SCIENTISTS, BUSINESS, AND THE STATE, 1890-1960

Patrick J. McGrath

The University of North Carolina Press

Chapel Hill and London

Designed by April Leidig-Higgins
Set in Minion by Keystone Typesetting, Inc.
Manufactured in the United States of America

The paper in this book meets the guidelines for
permanence and durability of the Committee on
Production Guidelines for Book Longevity of the
Council on Library Resources.

Library of Congress Cataloging-in-Publication Data
McGrath, Patrick J. (Patrick Joseph), 1961–
Scientists, business, and the state, 1890–1960 /
Patrick J. McGrath.
p. cm. — (The Luther Hartwell Hodges series on
business, society, and the state)
Includes bibliographical references and index.
ISBN 0-8078-2655-3 (cloth: alk. paper)
1. Science and state—United States—History—
19th century. 2. Science and state—United States—
History—20th century. I. Title. II. Series.
Q127.U5 M37 2002   338.97306—dc21   2001035149

06  05  04  03  02   5  4  3  2  1

To my parents,
James P. McGrath and
Anne T. McGrath

# CONTENTS

ACKNOWLEDGMENTS

I was extremely fortunate to have been trained as a historian at New York University, where I was generously supported intellectually and financially during my course work and while writing my dissertation. Without the extraordinary support of my thesis adviser, Thomas Bender, this project would never have been completed. He read two complete drafts of this book and provided insightful criticisms and suggestions. I am deeply in his debt for this and many other forms of assistance. Special thanks go to several other faculty members at NYU, especially Paul Mattingly, whose course Ideology and Social Change inspired me to take the plunge and become a historian. Danny Walkowitz, Carl Prince, Patricia Bonomi, and Marilyn Young also provided crucial support.

Thanks to my dissertation defense committee, which consisted of Professors Bender, Mattingly, and Young as well as Nathan Reingold of the Smithsonian Institution and Alan Brinkley of Columbia University. They all provided invaluable suggestions for turning that thesis into the present book, as did Professor Jessica Wang of UCLA. But I am solely responsible for any errors of fact or interpretation that remain.

The staffs at the many libraries and archives I visited deserve special thanks, not just for permission to use their materials (which I hereby gratefully acknowledge) but also for their extraordinary efforts above and beyond the call of duty. The people at the Library of Congress, the National Archives, the Rockefeller Archives, the New York Public Library, Harvard University, MIT, AT&T, and the Dwight D. Eisenhower and Harry S. Truman Presidential Libraries were all extremely helpful.

My thanks to Lewis Bateman, David Perry, Charles Grench, and the entire editorial and production staff at the University of North Carolina Press for their work in bringing this book to fruition.

I have many friends to thank for their support and encouragement over the long process of writing this book, especially Richard E. Mooney, Bill Alexander, and the gang at the Citicorp Center, as well as Lourdes Rodriguez, Rosa Taveras, Kaylee Rodriguez, Emma Segarra, Kelvin Rodriguez, Thomas Biegacki, John Rieger, Alice Fahs, Marc Aronson, Kathleen Hulser, Glenn Horowitz, Don Daniels, Gloria Price, Poppy Quattlebaum, Brian T. McGovern, and Rafik Cezanne.

My greatest debt is, of course, to my family, especially my parents, James and Anne McGrath, who taught me lifelong lessons by their examples of hard work, decency, and dedication. This book would not have been possible without them. I owe them everything, and I gratefully dedicate this book to them.

SCIENTISTS,
BUSINESS,
AND THE STATE,
1890–1960

Scientists allied themselves with America's corporate, political, and military elites between 1890 and 1960, and they did so not just to improve their professional standing and win more money for research but for political reasons as well. They wanted to use their expertise and their new institutional connections to effect a transformation of American political culture. They succeeded, but not in ways that all scientists envisioned or agreed upon. By 1960 America's governing ideology was organized around two key goals: increasing prosperity and enhancing military strength, and the centrality of consumerism and militarism owed much to the ideas and institutions which leading scientists created and used in the corporate and political sectors over the preceding decades.

Beginning in the late nineteenth and early twentieth centuries, scientists stressed their own importance to expanding the scope and effectiveness of national institutions. By doing so, they helped create a new kind of national authority, one that was rooted in the large corporations, the universities, the professions, and the federal government. They also helped foster a new conception of how social change occurred in American life. By stressing the evolutionary and transformative quality of their creations, they argued that scientific innovation was the central force driving American progress, prosperity, and national security. This new notion of social change led some scientists and intellectuals to formulate a new definition of democracy itself, one that rejected the old model of adversarial party politics in favor of a vision of a harmonious, classless meritocracy in which all members of the society would enjoy an improved standard of living provided by the corporate culture, and the talented would have more opportunities to enter the professional world.

In short, the leaders of American science in collaboration with corporate, political, academic, and later military elites created a new governing ideology in America in the years from the early twentieth century to the 1960s, one that relocated social authority away from local elites and political parties and toward a new national class of interrelated elite institutions in the public and private sectors. But this process was characterized by conflict every step of the way: between scientists and their powerful institutional collaborators and among the scientists themselves. This book is a study of those conflicts and the ideological, professional, and political factors that were under dispute.

It is also a study of the larger cultural consensus within which those conflicts occurred. The new governing ideology and the new conception of political authority that I examine were fashioned in conjunction with the rise of a powerful corporate economy. New monopolistic businesses used science and professionalism to legitimate their power, and scientists and professionals in turn used this new corporate system to enhance their status and achieve greater opportunities for wealth, power, and prestige.

Corporate elites and scientists engaged in collaboration and mutual exploitation, and in Chapter 1 I examine the cultural terms on which that exchange was carried out. It was hardly a placid relationship. It had built-in ideological tensions, which I describe in Chapter 2, and these tensions were largely responsible—as I argue in Chapters 3 and 4—for leading a group of scientific administrators in the late 1930s and early 1940s to forge a partnership with the military and the state which had profound consequences for American politics and culture.

The organizational and ideological terms on which scientists collaborated with the state were, I argue in Chapters 5 and 6, continuations of the terms on which they had been collaborating with the corporate sector over the preceding generation. Many of the same problems that existed in the corporate world reappeared in the relations between scientists and the military and political elite: who would make decisions about the kind of scientific work performed? Who would control its uses? Would scientists be equal partners and collaborators or mere functionaries and technicians?

Paradoxically, the very ideas that scientific elites used to create their collaborative relationships with corporate and military elites would be used by other political and scientific leaders to subordinate the scientists and to make it brutally clear, as in the case of J. Robert Oppenheimer, that scientists were regarded as mere "technicians," not governing partners. The very same ideas and institutional strategies that made scientists partners with the powerful also led to their abject humiliation and to the rise of a militaristic conception not only of American science but also of American political culture.

That paradox points to another that I would like to offer as one of the themes of this book: a new system of elite governance emerged that allowed scientists to enjoy greater influence in public life, but individual scientists continually found themselves receding from, or being pushed out of, the limelight and back into the obscure shadows of the organizational and professional structures from which they emerged. Consequently, historians have rarely taken the political ideas of scientists seriously. Specialization among historians partly accounts for this situation. Historical attention to

the emergence and growth of American science has been artificially seg-
mented from the scholarship on American intellect, politics, and culture.
The periods of specialization that prevail in the historical profession have
constricted understanding about the connections between science and the
larger stream of political life.

Histories of particular scientific institutions have been written, and many
works cover science in the nineteenth century, the pre- and postwar years,
and the Cold War.[1] Many of these studies have been excellent in excavating
the internal history of American science, but they have failed to place these
discrete stories in the larger context of the transformation of American
political culture from the 1890s to the 1960s. The use of science was central
to that transformation, and scientists and their institutions did not simply
respond to events in the political world but actively helped shape those
events. And they shaped them in the confines of an existing and changing
cultural setting.

The America of 1890 was still a political economy dominated by "courts
and parties," as the scholars Morton Keller and Stephen Skowronek observed.[2] New scientific innovations emerging from urban technical communities and financed by New York's investment elites led to the creation of
large, science-based corporate monopolies in the electrical and chemical
fields. These monopolies were faced with the difficult task of effectively
managing their far-flung enterprises and the even trickier task of gaining
public legitimacy in a political culture that was hostile to monopoly power
and badly divided along class and racial lines.

The leaders of these companies, in conjunction with intellectual and
scientific elites who had their own professional and political agendas, fashioned an ideology in which the process of corporate production and innovation was elevated to a central place in American culture. The production
of abundance through corporate science and expert management was offered as the justification for this new political economy. This emerging
ideology continued to shape American science during World War II and the
Cold War period, but its terms were continually contested. Ideological conflicts over the meaning and purpose of American science were simply continued in new forms and in different contexts. Issues of militarism and
national security made American science more intensely politicized than it
had been, and the explosion of state support for scientific research represented both a radical departure and a continuation of the ideological issues
that had accompanied American science in the prewar decades.

Twentieth-century Americans were still trying to work out a set of ideological and cultural values for the new economic world in which they found

themselves. The people who achieved dominant positions in the scientific profession in the 1940s and 1950s tried to answer those unresolved questions: What was the role of the professions, the state, and the business elite in a complex, technically driven national economy? How were problems of class relations, social mobility, and political power to be handled in this new setting? What were the new rules of the game? What was to be the new way of things?

Scientists and scientific administrators such as Frank Jewett, Karl Compton, David Lilienthal, Vannevar Bush, and James Conant sought to create an institutional partnership that included business, the professions, the military, and the political elite through which (they hoped) the very act of cooperation would resolve all problems related to national security, economic progress, and social stability.

But the associational visions of these men, which were derived from the dreams of social harmony expounded by Herbert Hoover and other Progressives, were supplanted by the ideas of a more militaristic faction of scientific leaders. Scientists and administrators such as Edward Teller, Lewis Strauss, and Ernest Lawrence, with their full-throated militarism and anti-communism, pushed American scientists and their institutions toward a nearly complete and subservient devotion to American military interests. The visions of science propounded by these men took the ideologies of preceding generations of scientific leaders and manipulated them through a set of political and institutional actions into a new ideology: scientific militarism. One vision of American science won out over another, with important consequences for American political culture. The public role of scientists receded further and further into the institutions of corporate science and the state and away from the people, whose volatility was feared and whose approbation was solicited. It made, by 1960, for a very strange political culture in which scientists and their institutions were central to America's governing structure, and yet they were largely invisible to the public.

I have chosen to focus my attention primarily on the scientific elites and on the way political change was refracted through and caused by their arguments with each other. It makes sense to study such an elite group for several reasons. The first is a selfish one: these people constitute a manageable set of historical subjects. But they are also the right people to be studying. They were the ones who created the system of relations between science, the state, the universities, and the corporate sector. They certainly did not control that system, but they created the ideas that were its ligaments, holding it together.

This story of the interplay between ideas and institutional politics, between professional cultures and political factions, has its own intrinsic historical fascination. But there is a larger purpose to this effort as well. I want to illuminate something of the way in which historical change itself was occurring in twentieth-century American political culture. The institutional complexity of American life in that period has always somewhat confounded scholars, particularly with regard to the interplay between ideas and institutional change. What was driving the process of change in these years?[3]

My story does not provide an answer in terms of any one class or any one theory. Instead I show how the ideological conflicts and the clashing professional and institutional cultures of scientists propelled and shaped social change. I am trying to combine intellectual history with institutional and political history because I am concerned with discovering how these overlapping processes resulted in the ideologies that governed political life in twentieth-century America.

I have operated on the assumption that ideological change is a staggered process: the old ways of thinking and acting do not suddenly depart and conveniently make way for the new. The new ways rise out of the problems, the contradictions, the weaknesses, and also the strengths of the old. But neither ideas nor institutions, and certainly not science and technology, are ever conceded dominance in this narrative. My focus is on the way specific people, in specific times and places, *used* ideas and institutions.

I must say at the outset that in the events I discuss my book does not represent an exploration of untouched historical ground. The debates over the creation of the atomic and hydrogen bombs, the organization of postwar scientific institutions, and the controversies over arms control and nuclear proliferation in the 1950s have been examined by other scholars. But their emphasis has been different from mine. I have taken some familiar episodes and some not so familiar and cast them in a larger chronological and cultural framework than most histories of American science or Cold War militarism have used. Rather than examining the military or diplomatic or professional implications of these events, I am principally concerned with the ways in which they shaped the boundaries of scientists' political and professional ideas and how those ideas shaped American politics generally. The different perspective I provide and the different questions I raise about these events offer a deeper level of understanding than has been provided by the narrow conceptual and chronological constraints of the existing literature on the history of American science.[4]

The people who created what I call corporate science and state science sought to achieve an ideological compromise between the values of their profession, the values of America's governing institutions, and the political and cultural values of the public, as they understood it. They tried to demonstrate how the values they prized could be effected by working within a system that would absorb and in some cases pervert their values. By the 1960s, the point at which, roughly, my story breaks off, the system they helped establish and the militant Cold War ideology that sustained it were firmly fixed at the center of America's political culture.

But the triumph of scientific militarism was not inevitable. It was not the result of the "evolution" of science or some mystical triumph of the machine or the result of sheer bureaucratic or technical momentum. It was the result of certain ideological, institutional, and political choices made by actual human beings. This book is about those people, their choices, and the consequences that followed from their actions.

## CHAPTER 1

## SCIENCE AND THE CRISES OF
## THE NEW CORPORATE ORDER

Scientific elites in the late nineteenth and early twentieth centuries continually spoke about the ability of science to resolve conflict and create harmony. They had excellent reasons for hoping so because the America of that period was a society badly split by racial, ethnic, and class conflict.

The emergence of new science-based corporate monopolies had created wide disparities in wealth and new relations between workers and their employers that often led to extreme exploitation and equally extreme violence. The social fabric may not have been as badly frayed as it was in the 1850s, but memories of a convulsive Civil War were easily triggered by the violence and bloodshed of the summer of 1877, by the Haymarket incident, and by the Pullman Strike and the political violence that seemed endemic by the turn of the century. Many middle-class and upper-class Americans were eager for some strategy for eliminating social conflict and restoring stable relations between workers and employers. They wanted to reconcile traditional American notions of democracy, opportunity, and self-reliance with a new economic order that made a mockery of the Lincolnian notion that any man could aspire to become his own economic master.[1]

Science offered a way. Many Americans turned to it as an instrument for restoring order and stability and reconciling traditional values with current realities. For some conservatives it could be used to justify social inequities in terms of the prevailing values of nineteenth-century America.[2] Thus in a culture that flattered itself that its history was the unfolding of a divinely sanctioned plan, monopoly power and gross economic inequality were justified as the consequence of Social Darwinian "evolution." The "failures" of the poor, the immigrant, and the nonwhite were explained by resorting to the prevailing racist categories of thought: the poor were poor because of their racial inferiority or because of defects in their unreformable genes. The rich and powerful, by contrast, were deemed paragons of individualism, the Protestant work ethic, and good blood.[3] But this intellectual "patchwork" could not last. Something more substantial and permanent was needed to close the tears in the fabric of American society. The political violence and turbulence of the era could not be argued away by invoking Anglo-Saxon superiority.[4]

The first group to use science in support of a new social vision were academic social scientists who wanted for both professional and ideological reasons to root elite authority in America on scientific knowledge. By doing so they intended not only to establish the importance of their disciplines but to make the troubling world of the late nineteenth century comprehensible and politically palatable.[5] Unable to control change, elites sought instead to control its definitions and its meanings.

The social science enterprise in late nineteenth- and early twentieth-century America was an attempt to explain social change as a "natural" process that would work itself out to create a stable social order. The new social science disciplines developed an analysis that depicted the interdependence, mobility, and pluralism of American life as forces that were as natural and predictable as forces of nature and would always result in a stable capitalist society.[6]

Under the rubric of academic social science, scientific ideas like evolution could be used as an argument for Progressive reform as well as for reaction. This crucial idea was first grasped by Richard Hofstadter back in 1944, when he noted that liberals like Lester Ward were using the same ideas to argue for activist reform that William Graham Sumner had used in support of the creed of rugged individualism.[7] Scientific ideas could be, and were, used by many different groups for very different reasons.

This idea of using science to stabilize and harmonize a troubled society spread slowly from academia to the intellectual and business communities. Corporate leaders in the new science-based industries used an idealized vision of science to advance their own cultural and political goals. They would draw on Progressive Era and social scientific ideas about evolution, economic interdependence, and the authority of expert knowledge to create a new conception of American political culture.

And what would the new, cooperative, scientifically based culture look like once the older, adversarial model had been put to rest? Americans in the early twentieth century were divided over how to proceed at the cultural crossroad they had reached. Morton Keller was writing of the Gilded Age, but his words apply as well to the Progressive period. It was "a time of intense conflict between old values and the pressures generated by massive change."[8] Various business elites grasped at science as a means of refashioning the social order. But how exactly would science be used? How could it reconcile social conflict or preserve the older, nineteenth-century social order in the midst of a rapidly changing political economy?

One way of answering these questions is by comparing the experiences of two very different businessmen and scientific institution-builders in this period: Henry R. Towne and Theodore Vail, the latter a creator of a powerful monopoly, the former the benefactor of a failed and obscure museum. The different experiences of these two men and their institutions express the crucial historical change that was taking place in this period, namely the emergence of a political and cultural order rooted in new national institutions, in place of the local and regional economies that had been dominated by local business and political elites during the nineteenth century.

Vail, the president of the American Telephone and Telegraph Company (AT&T), succeeded in creating a national telephone monopoly by blending older values with new institutional forms, by tapping into powerful cultural notions of technological progress and enhanced material prosperity, and by establishing institutional connections with the scientific profession. Towne, an independent businessman and engineer, tried and failed to establish a museum dedicated to the mechanical arts in America. Both men tried to use ideas about science to achieve political stability. But the terms of the stability they sought were strikingly different.

Towne hoped his museum would foster respect and loyalty for the achievements of the business class of which he had been a part, and he hoped it would serve as a cultural bulwark against the political storms of his era. He was the owner of the Yale and Towne Manufacturing Company and an early champion of scientific management, as well as a leading figure in the postbellum New York business world. The museum he endowed had initially been chartered by a group of patrician engineers and business leaders in 1914 and was originally named the Museum of the Peaceful Arts. The founders were almost all men born in the mid-nineteenth century, and their intentions in creating the institution were pedagogic and moralistic. They were very much in the mold of the prominent citizens who helped form the late nineteenth-century museums in such major cities as Boston and New York.[9]

They wanted to teach the working class, whom they hoped would be the principal audience at the museum, about the accomplishments and presumably the wisdom of the employing class. They also hoped the duly impressed audience of laborers would become more knowledgeable and cooperative about their place in an interdependent, technological, and industrial urban economy. The endowment of this museum was the culmination of a career in which Towne had sought in various ways literally to codify stable class relations. He was an early proponent of "gain sharing" plans that he used in his own business and hoped would be widely adopted. He proposed a written contract between workers and owners in the railroad industry in 1916 as a way of ending a bitter strike and preventing future ones. He also advocated the creation of governmental bodies staffed by experts to handle problems like the tariff on a "scientific" rather than a political basis. He wanted to use scientific ideas and expert knowledge to render permanent and stable the existing class system in America.[10]

After Towne's death this agenda was continued in his museum, which got started in modest quarters in the Scientific American Building in New York under the direction of Towne's friend George Kunz, the president of Tiffany and Company. Kunz described the museum to potential donors as a place that would "help create intelligent and productive workers, and an intelligent workman who has an interest in his profession is rarely ever lazy or out of employment—and often it is unemployment that leads men to commit crime." This notion of reforming the habits of workmen was redolent of nineteenth-century programs of moral reform and uplift. It was expressed in many forms and contexts, such as in the ideas of Massachusetts Institute of Technology (MIT) founder William Barton Rogers, who hoped his school would "influence the morality and the intelligence of the industrial classes." And, of course, it is evident in the ideas of Frederick Winslow Taylor, the great bogeyman of all post-1960s labor historians, who is mistakenly seen as the inaugurator of a new, authoritarian brand of capitalism, whereas he, like Towne, Kunz, and all the other gentlemen reformers who wanted to shape up the working classes, was about to be brushed aside in favor of a new set of strategies employed by business elites like Vail.[11]

The challenge presented to Kunz in his own museum is illustrative of this new strategy. The static, pedagogical conception of class relations embodied by the museum was contested by Kunz's trustees such as Charles Gwynne, the vice-president of the New York State Chamber of Commerce, who felt the focus of the museum should not be on educating the public but on promoting the image of American corporations. Gwynne hoped to solicit funds from U.S. Steel and Standard Oil, with the view that they would create "industrial exhibits." In promoting the museum, Gwynne also spoke of educational goals, but he was not concerned about educating patrons to be better workers. His goals were more abstract, and the agenda of corporations was rendered more faint and obscure. He wanted to show individual workers "the relationship of his industry to him as well as his relationship to that industry; he must be given perspective necessary to realize the interdependence of that industry and other industries to the national prosperity."[12]

He had no desire to have the museum educate and reform the visitors in the manner Kunz described. Industrial museums "of the new age," Gwynne wrote, "are remarkable in taking people as they find them." They were built "to survive all manner of misunderstanding as long as visitors just come, no matter what they come for." Once they were there, he was confident they would receive the messages he wanted to convey about corporate power.[13] Towne and Kunz sought to use their institution to stabilize a particular set of class relations between employers and workers; Gwynne was more interested in establishing in the public mind the centrality of corporate institutions in a prosperous, scientific, and interdependent economy. It was Gwynne's view that ultimately prevailed in the museum. By the 1930s, when the museum fell under the control of a group of corporate executives led by AT&T's Frank Jewett, it became essentially a showroom for corporate products.

Vail's success at AT&T, by contrast, was the result of his ability to create a more abstract, fluid sense of political relations in an economy increasingly dominated by large, monopolistic companies. He successfully evoked the sense of interdependence that Gwynne spoke about, and he combined the American enthusiasm for invention and technical progress with the image of a progressive, national corporation that was at the vanguard of American improvement and prosperity. Above all, he was able to transform AT&T in a way that provided new professional opportunities for engineers and scientists, as well as new social opportunities for American consumers. Rather than simply presenting elite values to the public in a didactic manner, Vail offered the appearance of a dynamic national force that was transforming the country in a positive way.

He became the head of AT&T in 1907, after a shakeup in the company that led to the removal of several of the old Boston elites that had dominated the company's board. They were replaced by members installed at the behest of J. P. Morgan, whose infusions of capital had allowed the company to weather the Panic of 1907. Upon taking over, Vail was intent on both consolidating the monopoly he managed and simultaneously making it seem to be a benign force. He developed an oft-repeated slogan that he offered as the company creed: "One Policy—One System—Universal Service." He successfully deflected the considerable public hostility that had built up against AT&T by softening the bare-knuckled competitive tactics that were used against the smaller "independents." He discontinued the policy of refusing to connect their lines with AT&T lines in adjoining regions. Rather than trying to drive the independents out of business, Vail planned simply to absorb as many as possible into the regional companies. The additions to the system would then strengthen Vail's public depiction of the company as a "natural monopoly."[14]

He also made scientific research a top priority. Earlier efforts at instituting basic scientific research had been thwarted by the cost-conscious directors. Until 1910 the company had relied on contractual relations with outside inventors such as Michael Pupin to provide its technical innovations. To ensure that the company had the equipment it needed for its expansion, scientific discovery and mechanical invention would henceforth have to be carried out under the same institutional roof. The cultivation of internal invention also had larger corporate and even political consequences: developing its own inventions would get the company out of the messy business of fighting patent challenges in the courts. In 1891, a Bell lawyer noted that the hundreds of patent suits the company defended against independent inventors made its monopoly "more profitable and more controlling— and more generally hated—than any ever given by any patent."[15]

To achieve Vail's economic and public relations goals, the company had to develop the capability for transcontinental telephony. It had to create a national wire network that removed competitors and made the Bell System seem essential and irreplaceable. The pursuit of this goal led, in 1908, to the consolidation of the company's disparate research efforts. Vail appointed John J. Carty director of research. He wanted to hire scientists and engineers who could engage not just in "applied" technical work but in basic scientific research that might lead to fundamental technological breakthroughs.[16]

This is an example of how companies like AT&T, General Electric, and Du Pont that were starting their own scientific laboratories around this time created a new demand for more scientific and especially engineering expertise in America. These corporate changes would be the catalyst, as well as the beneficiary, of a transformation of the scientific professions in these early decades of the twentieth century. For their part, America's engineers were determined to improve their status relative to the nation's other dominant professional groups such as lawyers, business executives, and politicians.[17]

An important shift in the ideology of the engineers occurred in the 1910s and 1920s. Earlier engineers, in the mold of Henry Towne, had thought it possible to carry out their professional and political goals from an autonomous, independent professional position in American culture. One engineer succinctly expressed this sentiment by envisioning the implementation of the golden rule by use of the slide rule. But as engineers of all stripes became increasingly affiliated with large corporate businesses, their ideology was expressed as a set of principles that would be realized in conjunction with American business. George F. Swain put this new creed bluntly: "The engineer is a businessman, for engineering is business and business is engineering."[18]

This fusion of the professional identity of engineer-businessmen, or corporate scientists, with the interests of large corporate organizations was vividly expressed in the new corporate research laboratories, and preeminent among those was AT&T. When the company engineers took up Vail's project, they were developing their own distinctive professional identity, as well as strengthening their employer. This was one of the keys to AT&T's success and why it was politically able to sustain its monopoly position. Its actions had cultural ripple effects that led people outside of the institution—such as America's scientific and engineering communities—to want to see it succeed.

But this is by no means to suggest that the fostering of engineering and scientific professionalism within the company was automatic or easy. Far from it. Engineers and scientists had to fight to establish their legitimacy and their value within the corporation. In spite of the eagerness of the new management to enhance technological productivity, the reorganized research department was still operating under some constraints. The Morgan-appointed directors on the board issued strict instructions about cost control and expected any efforts in basic research to pay off. Many fledgling corporate research labs labored under similar constraints. In 1938, Frank Jewett, who was a member of Carty's research organization, assessed the early years of industrial research: "One thing stands out in every case with which I am familiar, namely, the hard internal sledding of the initial years." The research-minded scientists were looked on by the "practical" business managers of the organizations as " 'impractical' theorists," and the "first big job was to sell the new idea" and win the "confidence and respect" of the company's managers.[19]

Engaging in such salesmanship forced corporate scientists like Jewett and Carty to craft a professional identity that stressed their importance to corporate and even national progress. Carty was very adroit at such salesmanship. He carried out highly public demonstrations of the labs' accomplishments and succeeded in winning the support of the board for more investment in research. In the public press, Carty cast the transcontinental telephone project as a patriotic quest for national unity. On a trip to some northwestern affiliates, he passed through San Francisco, where plans were under way for the Pan American Exposition that was to take place in 1915. He boldly, perhaps recklessly, promised that AT&T's transcontinental network would be completed in time to be demonstrated at the exposition. He also explained the inspiration for this effort and omitted, predictably, any business motives for it. Instead, he said that his travels through the Northwest had impressed upon him the "desolation and loneliness" of the area. Passing through downtown Seattle, he had come upon a statue of William Seward and became filled with patriotic sentiments. He vowed to achieve through technology what the Unionists of the 1860s had tried to attain through the violence of civil war: the unification of the country.[20] The truth, however, was that achieving a national wire telephone system was especially urgent because some of AT&T's competitors were indicating the first stirrings of wireless telephony.[21]

Fortunately for Carty, his boasts were made good. A team of academically trained physicists was needed to tackle the problem and Jewett convinced Carty to hire a cadre of physicists who could work on the system. Jewett's old friend Robert Millikan served as a recruiter of potential candidates and was responsible for sending Jewett young scientists including H. H. Arnold, John Mills, H. W. Nichols, and K. K. Darrow, men not only who would play an important role in the immediate context of the transcontinental project but who would also rise to important executive positions in the company.[22] Here again the creation of interinstitutional links would serve to strengthen the AT&T monopoly.

In January 1915, an aged Alexander Graham Bell was enlisted to come to New York and speak over the phone to his slightly less aged assistant, Thomas Watson, so they could recreate their famous first telephone conversation, this time over a distance of three thousand miles. The event was highly publicized and attended by company directors in New York so that they might see for themselves the big payoff for their investment in research.[23] The lab was also able to come up with workable, long-distance wireless telephony, and Carty also made sure that the press was present to record the accomplishment. On September 29, 1915, he traveled to Mare Island, off San Francisco, to receive a call from Vail in New York. The call was transmitted along AT&T's wire network to San Francisco, where the transmission was automatically switched to a radio transmitter and broadcast to Mare Island. The next day the line of communication was extended to Pearl Harbor, Hawaii, where a company engineer received a transmission that originated from the Arlington naval station.

Theodore Vail grasped the usefulness of these achievements for promoting the company's economic and ideological concerns. He would use the scientists for corporate purposes, just as the scientists were using the corporation for their own professional purposes. He seized upon Carty's successes to craft a public image of his monopoly that depicted AT&T as an instrument of social and cultural, not just technological, progress. In an October 17, 1915, piece in the *New York Times Sunday Magazine*, Vail reiterated the irreplaceable nature of wire telephony. He also expounded on a favorite theme among AT&T publicists: the harmonizing effects of the company's technology.[24]

The transcontinental system would facilitate greater cooperation, Vail argued, thus inevitably bringing about social harmony. The telephone, he said, "reduced the whole United States, potentially, to the size of a community." It was also the preeminent method of communicating within this "community." "You can't argue by mail," Vail claimed, "and it is through argument that conclusions of great moment are reached. Discussion—that's the thing."[25] But a none too deep exploration into Vail's ideas indicates that consensus was the thing. "I feel confident," he said, "that this country will never again have any difficulties which cannot be settled by discussions and mutual concessions." If a "free interchange of ideas and thought" could be achieved, then "a common course of action will be found to rectify most troubles." He described the "free interchange" of ideas through the telephone as a vehicle for greater public "education." And it is "through education of the people," he claimed, "[that] we will achieve non-partisanship." A "non-partisan" public philosophy, he argued, was one that led to greater wisdom about the regulation of telephone rates.[26]

Vail concluded his "interview" by offering a justification not only for his company but for monopoly in general (although he dared not speak its name). Here his words seem practically to be lifted from Walter Lippmann's *Drift and Mastery*: "We must realize that we can't do big things with little instrumentalities. . . . The bigger the situation, the bigger the opportunity for all; but for big situations, big men must be found who can think big thoughts, interpret them in big ways to the public, and thus accomplish big results."[27]

The corporate scientists were not quite the "big men" Vail envisioned. But they were certainly working in big institutions, planning their work in accordance with the larger economic interests of their corporate directors. They saw themselves as professionals existing between the worlds of business and academia, pursuing the most admirable qualities of both realms. They were scholarly businessmen and practical academics.[28]

The writings of Vail and Carty (and, as we will see in the next chapter, the writings of Bell Labs president Frank Jewett) expressed a conscious awareness of new cultural as well as economic opportunities being created, of new professional paths being forged, that would create a new society. AT&T's leaders were simultaneously recasting their public image by devising new professional opportunities and new labor relations in the company so that, in a strange way, innovation and reaction existed symbiotically.[29] While workers and middle-class groups inside and outside the company embraced those new products and professional opportunities enthusiastically, the notion of the primacy of corporate capitalist power was sustained.

Vail was blurring the line between employers and workers in his ideology and devising a political language that rhetorically equated the highest and lowest members of the corporation. He was able to endorse the radical notion that "labor should manage all industry and get all the profits" because managerial "labor" was no less a form of labor than manual work. But of course the managerial "laborers" controlled the money and the decision-making powers of the company. Compare this institutional and ideological strategy with the writings of Towne and his colleagues in his museum enterprise. The latter were a group of men trying, almost forlornly, to keep alive a dying world. They were trying to sustain a static, local vision of cultural authority and social change in which progress and innovation flowed from the efforts of the elite entrepreneurs who ran society, and a diligent and deferential working class placidly played its subordinate role in this ideological scheme.[30]

Vail and his corporate science colleagues, by contrast, were suggesting a different model of both social authority and social change, one that was resonating more powerfully in Progressive Era America. A comparison between the rhetoric of the corporate scientists and some of the leading Progressives is worth considering to illuminate the ways Progressive political thought was deeply influenced by the major institutional changes that were occurring in the American economy and how the leaders of those new institutions, in turn, realized the need to legitimate their monopoly status by resorting to progressive ideas (big and little "p"), which they hoped would capture the public imagination and protect their new political status.

Liberal thought and the political culture were changing together, although certainly not in any lockstep fashion. But there was an important interplay at work in this period, one that would continue throughout the ensuing decades as well. Academics, intellectuals, and corporate promoters were creating a body of ideas that came to constitute a vision of reality—an ideology—that made it possible for some political goals to be achieved while blocking others. Corporate monopolies such as AT&T, General Electric, and Standard Oil were forcing a radical redefinition of the republican tradition, one that subsumed notions of independence and individualism into a corporate context and defined freedom in terms of material consumption and the opportunities and rewards offered by the corporate system. The proponents of this ideology were trying to answer questions that had been disturbingly persistent throughout the wrenching decades of the Gilded Age. How could the values of economic independence, self-reliance, and social stability be preserved in an economy dominated by large corporate institutions and marked by widespread misery and conflict?[31]

Three ideas proved especially important in answering these questions and effecting what might be called the ideological reconstruction of American capitalism. The first and most prominent was an emphasis on evolutionary development. The growth of large corporations was depicted as an inevitable and benign process that Americans need not fear. The second was that the new corporate economy provided opportunities for the exercise of expertise and intellectual skill that made it appealing to an emerging middle class of educated professionals. The third idea was that both the evolutionary growth of corporations and the emergence of experts managing this system would redound to the benefit of the general public, the "consumers." A new, stable political order would thus emerge under the aegis of the corporate scientific order.

The idea of evolutionary development had the oldest intellectual pedigree. American attitudes about science and technology had long been connected with the belief in a Divine Providence that oversaw American national growth. Technological innovation had also long been seen as a vindication of the virtues of American republicanism.[32] The strength and longevity of this idea help explain the immense popularity of Edward Bellamy's *Looking Backward*, a laughable work of art but a cultural artifact that captured the anxieties and hopes of Americans in the midst of the political crisis of the 1880s and 1890s. The very brightness of its utopian vision suggests the disturbing bleakness of the real-world context against which Bellamy was writing.

The novel's protagonist, Julian West, described his reaction to being transported into the late twenty-first century in terms that many of Bellamy's contemporaries might have applied to their own perceptions of America's violently rapid industrial revolution: "In my mind all had broken loose, habits of feeling, associations of thought, ideas of persons and things, all had dissolved and lost coherence and were seething together in apparently irretrievable chaos."[33] But the Boston of 2087 was anything but incoherent and chaotic. It was a utopia. It was a world in which the state had taken control of all means of production and had established an "industrial army." Money, deprivation, and political conflict had all been abolished. Bellamy made it clear that this utopia had come about as a result of Americans wisely deferring to the evolutionary process that was under way in their society. He had one of his characters explain how American society solved one of the bitterest social and political problems of the 1880s, the "labor question."

It was an explanation that entirely ignored political action: "As no such thing as the labor question is known nowadays . . . and there is no way in which it could arise, I suppose we may claim to have solved it. . . . It may be said to have solved itself. The solution came as the result of a process of industrial evolution which could not have terminated otherwise. All that society had to do was to recognize and cooperate with that evolution, when its tendency had become unmistakable." Another character explained in a sermon that the changes also came about with great rapidity. "In the time of one generation men laid aside the social traditions and practices of barbarians, and assumed a social order worthy of human beings." They ceased being "predatory in their habits, they became coworkers, and found in fraternity, at once, the science of wealth and happiness."[34]

Like Charles Dickens's Mr. Scrooge, Americans just "woke up" and changed. They suddenly realized, without any coercion or conflict, that they had been living their lives destructively and they set about making things better. They had been converted. The novel is remarkable in its repeated emphasis on the irrelevance of human action to the changes it describes. Here is where Bellamy would differ most strikingly from the works of such later Progressives as Walter Lippmann and Herbert Croly: he made no provision for the role of experts. The new order was made to appear as if it were produced by nothing more assertive and tangible than sentiment. Its smooth operation was only slightly attributable to human influence. "The functionaries at Washington to whom it is trusted," a character declares, need be "nothing more than men of fair abilities to discharge it to the entire satisfaction of the nation. The machine they direct is indeed a vast one, but so logical in its principles and direct and simple in its workings, that it all but runs itself; and nobody but a fool could derange it."[35]

*Looking Backward* expressed a deep yearning for the creation of a social system without conflict. It also reflected the ideas of the new social scientists that change was a natural process that could be understood—and deferred to—with trained intelligence. Bellamy offered a vision of a polity in which the distinction between leaders and led, powerful and powerless, had been transmuted into an administrative problem. If the commercial success of Bellamy's book and if the hundreds of "Nationalist" political clubs it inspired are any indication, many Americans shared a similar, desperate urge to find a way out of a tumultuous system that brought immense material progress as well as greater conflict, exploitation, and misery.[36]

For all its commercial popularity, however, Bellamy's strict technological determinism failed as a political philosophy. Industrial evolution and the concentration of labor and capital did not by themselves bring about the great transformation he envisioned. But Americans (middle-class readers of Bellamyesque literature, at any rate) by no means abandoned the central idea of *Looking Backward*: that technological evolution offered hope of greater progress and greater social cohesion.

These ideas are evident in the speeches of AT&T's John J. Carty in the 1910s and early 1920s. They represent an effort at fitting the accomplishments of corporate scientists into the same paradigm of progress and evolution that Bellamy used. In a 1916 speech Carty noted that telephone technology "more than any other . . . reflects the genius of our people. The story of its wonderful development is a story of our own country. It is a story exclusively of American enterprise and progress." Modern Americans, living amid the wonders of technological innovation, were "pioneers in a new land" being refashioned and made anew by the scientists and the engineers, working together in powerful corporate monopolies. The pure scientists, Carty wrote, were nothing less than "the advance guard of civilization."[37]

In his 1920 essay "Science and the Industries," Carty gave full vent to the consumerist and utopian strands of popular thought that were flowing through the discourse of corporate science. "The possibilities of science," he assured the public, "are boundless." He urged the public to "no longer interpret life as a struggle among men for a limited share, where one man's gain must be another man's loss." They must instead "pay heed to the voice of the scientist and under his leadership join with their fellow men, all working together in controlling and utilizing the bountiful forces of nature." Like the early American pioneers on the frontier, modern Americans must temporarily endure hardships until they could be "buoyed up," as were their pioneering predecessors, "with that vision of vast natural resources which unfolded itself before their eyes." Modern Americans must "look about them through the eyes of modern science and . . . see that they too are pioneers, and in a world of wonders filled with boundless promise which will be realized by their children and their children's children and all of their generations in increasing measure."[38]

This is rhetoric that is beyond utopian. It almost sounds like the language of a child's fairy tale. The rhetoric of corporate science espoused by Carty offered a vision of corporate culture in which the company, like a benevolent parent, cared for nothing except the well-being and the satisfaction of its customers and workers. Of course, it wanted to make money too, but this was accomplished only to the extent that the customer was satisfied.

Carty had a self-interested reason for using this language and imagery. The professional ambitions of the corporate scientists meshed nicely with an emerging corporate creed about the centrality of consumerism and material abundance in American life. By stressing the importance of science to corporate expansion, corporate scientists were not simply selling themselves to the directors of their organizations; they were presenting themselves *and* their corporations together to the public as the crucial figures in the reorientation of America's political economy.

Carty, who liked to point out to business audiences that in addition to his scientific affiliations he was a member of the Chamber of Commerce, could also be frank and reassuring to those with a strictly business sensibility. In a 1924 speech, "Science and Business," he noted that corporate research labs "are organized on a strictly business basis, and the work conducted in them is directed to no other purpose than improving and extending and conducting in a more economical manner the service which we render to the public. The criterion which we apply to the work conducted in these laboratories is that of practical utility. Unless the work promises practical results it cannot and should not be continued." The fundamental question any corporate scientist should ask or be asked, Carty concluded, is, "Does this kind of scientific research pay?"[39]

On occasion, he could blend the utopian and practical strands of his thought together in the same speech. "Applied science does pay for itself," he assured the National Electric Light Association convention in 1929. "Your industry could not exist without it." The costs of producing electricity, he pointed out, had decreased fourfold since 1907. But however important it was to discuss "science in terms of money," Carty told his audience that "the value of science can best be measured in terms of human achievement, the mastery of the forces of nature, the elimination of poverty and disease, the prolongation of life, the advancement of learning, the growth of right living and sound thinking and of good understanding among men."[40]

Regarding any public role for scientists outside the lab, Carty's thinking never strayed far from the simple abstractions of Social Darwinism and Bellamy's evolutionary utopianism. He was, according to his colleague Frank Jewett, "a true disciple of Herbert Spencer." He spoke of the research lab as a "collective mind" and regarded the national telephone system he helped create as a "national nervous system, binding the people and the geographical units of the country together" so that it "might function more smoothly as a well-integrated organism and reach that higher goal which represents perfectly coordinated cooperative effort."[41]

If anything, Carty saw the public role of the corporate scientist as that of constructing a new mentality, in which all aspects of society were redefined in accordance with his evolutionary creed. "It is the function of the engineer," he said, "to provide for the extension of the spoken word by means of electrical systems of inter-communication which will serve to connect the nervous system of each unit of society with all of the others, thus providing an indispensable element in the structure of that great and powerful organism which it is believed will be the ultimate outcome of the miraculous evolution which society is to undergo." But this harmony, like the harmony achieved in *Looking Backward*, would occur automatically, simply as a result of the growth and "evolution" of corporate science. And though Carty was involved on the fringes of the psychiatric and eugenics movements of his day, he never envisioned any assertive role for scientists and engineers in the actual institutions of government.[42]

Carty's thoughts were not idiosyncratic or eccentric. They were highly representative of the rhetoric of corporate science employed by many corporations in the early twentieth century. The leaders of America's major urban department stores in this era such as John Wanamaker and Marshall Field espoused an alluring vision of abundance, opulence, and utopian satisfaction. Middle-class magazines and advertising likewise depicted American capitalism as something that existed solely to satisfy the material desires of individuals. This consumerist rhetoric not only abolished the old moralisms about thrift, restraint, and frugality, but it also abolished—or simply ignored—any discussion of politics and social conflict.[43]

Other large national monopolies began copying the strategies used by AT&T and stressed their companies' democratic character, progressive impact on society, and the social harmony they were able to achieve under the leadership of benevolent expertise that was free from regulatory or trade union interference.[44] Even a socialist like Charles Steinmetz, the famous General Electric engineer, stressed the benign and evolutionary qualities of his employer and other corporations and celebrated the ways in which they could accomplish a salutary transformation of American political life. "I would rather take my chances with the impersonal, huge industrial corporation than with the well-meaning individual employer," Steinmetz wrote. The corporation was a superior cultural form because it already had the attributes of a democratic institution. The managers provided a republican form of representation, and corporations employed variations on the referendum and recall initiatives that were dear to Progressives' hearts. Above all, it was an efficient institutional form that stressed cooperation. The goal of American political culture as a whole, he argued, was "to extend [the] methods of economic efficiency from the individual industrial corporation to the national organism as a whole." He went so far as to argue that the corporate form should be copied by municipal governments. Like Bellamy, Steinmetz believed that such an evolution would occur naturally as long as meddlesome and inappropriate regulations were not applied to the corporation. "Economic laws are laws of nature," he stated. They are "inevitable, and their defiance, whether by an individual or by a nation, means self-destruction." The laws of nature "are above man-made laws, and political law violating the laws of nature is void."[45]

Such ideas were also echoed outside the corporate lab by prominent academic scientists such as Robert Millikan, who evangelized on behalf of the cause of expanding basic scientific research. In 1919, Millikan said in a speech, "It is probable that the total possibilities of improvement of conditions through distribution [of wealth] are very limited, while possibilities of improvement through increases in production are incalculable." Only scientific research, he made clear, would bring about the necessary increases in production. Millikan was also a fervent advocate of educating the public in the rational, experimental methods of science. He hoped the public would embrace this mentality as an instrument for settling social and even international disputes.[46]

Steinmetz was a socialist. Millikan was a conservative Republican of the rugged individualist school. Yet they both espoused a body of political thought that transcended categories of left or right, conservative or radical. The corporation and its innovative energies was the central, dynamic, and unifying force in the new political order outlined in their writings. Its unifying power was especially important to the proponents of corporate science. The contentious striving of individuals or groups was no longer necessary. In an innovative and abundant economy, all interests would be united within and around a corporate system that was generating prosperity. The notion that America's political economy was a single, unified entity began to appear in the thought of leading national politicians. Herbert Hoover also saw the economy as "a single industrial organism," with engineers "standing midway in the conflict between capital and labor." America's actual corporate system, of course, was hardly an organism that ran itself by some benign, abstract process of evolution. One of the most remarkable things about late nineteenth- and early twentieth-century American capitalism was the emergence of a professionalized, managerial middle class, which was aligned with the new corporate bureaucracies and entrusted with their smooth and efficient operation.[47] Middle-class Americans and aspiring professionals wanted to use this new institutional mechanism for their own ends. But, again, what should those ends be?

This professional class found both its chronicler and its champion in Walter Lippmann, who tried to answer that question in his famous 1914 work *Drift and Mastery*. Lippmann saw the rise of this new corporate managerial class, which he assumed was infused with the spirit and methods of science, as the cultural triumph of scientific thinking. In later years these attributes would be expressed vividly in the ideas of people such as Vannevar Bush and David Lilienthal, whose writings contained strong echoes of *Drift and Mastery*. For them, the cultural application of science was not some narrow application of a set of axioms but an expression of an energetic and positive vision of American culture. But it was also a vision that sought to evade and obliterate political conflict and political thought entirely.[48]

This aversion to politics was evident in Lippmann's work. The new authority he spoke about was not one that was to be asserted; it was merely to be made manifest. Lippmann's experts would simply be managing a system that had somehow sprung into being fully formed. He did not examine the history or the political significance of the new corporate monopolies. He was too excited about the institutional and intellectual opportunities they posed to pause and reflect on the political character of the system. He was content to point out that a new corporate economy had evolved, and he urged scientifically minded experts to jump on the evolutionary bandwagon and begin managing it.

"The trusts," Lippmann wrote, "have created a demand for a new type of businessman—for a man whose motives resemble those of the applied scientist and whose responsibility is that of the public servant." This was a very important redefinition of corporate culture. Lippmann was rejecting what he regarded as the simple moralisms of the muckrakers who attacked big business. For him, the trusts were providing fertile ground for the exercise of a new type of American professionalism.[49] "The real news about business," Lippmann wrote, "is that it is being administered by men who are not profiteers." This was overdoing it, of course. But he was partially right. The members of the new managerial class were concerned with efficiency, and as a result Lippmann optimistically believed they were more likely to work in partnership with labor and not make war against it.[50]

This new managerial class was making a "science" of administration. It would grow in professionalism, in strength, and in public prominence, and eventually, he predicted, "private property will melt away; its functions will be taken over by the salaried men who direct them, by government commissions, by labor unions." Now, this was really overdoing it. But such language allows us to see how close his political analysis was to the language of Bellamy's utopian fiction.[51]

Both Lippmann and Bellamy found the solutions to the problems they posed from within the terms of that problem. Bellamy's proposed solution entailed an intensification of the growth of industrial enterprise. He believed that this would naturally develop into a rational system of cooperation and even benevolent regimentation. Likewise, Lippmann saw the problem he described as an opportunity. The political, moral, and spiritual drift he decried did not lead him to call for the dismantling of the industrial system that gave rise to this culture. He instead called for enterprising young men like himself to take up positions of managerial and intellectual influence and to offer their expertise as a new moral, political, and spiritual creed. Both Lippmann and Bellamy rejected radical or revolutionary solutions to the problems they described. They instead suggested, in the spirit of both academic social science and Vail's corporate rhetoric, that the hopeful aspects of the existing situation could be turned toward a positive solution, but without any fundamental alteration of the nation's political economy.

Even the most marked difference between Bellamy's and Lippmann's work, such as Lippmann's emphasis on the crucial, visible role that experts would play in guiding the system, proves to be not so very different from Bellamy's vision when we consider that Lippmann's experts would be operating in a political vacuum. There would be no one opposing the expert leaders he celebrated. Indeed, the only conflict he acknowledged in his book was the one suggested in the title: between those who would complacently drift and those who would exert mastery. But this difference was merely an aesthetic one.

He ridiculed, for example, Woodrow Wilson's "New Freedom": the nation of small, independent entrepreneurs that Wilson's creed envisioned was, Lippmann wrote disdainfully, "an unworthy dream. I submit that the intelligent men of my generation can find a better outlet for their energies than in making themselves the masters of little businesses. They have the vast opportunity of introducing order and purpose into the business world, of devising administrative methods by which great resources of the country can be operated on some thought-out plan."[52]

Lippmann's bold new men wanted action and power. But their actions and their ambitions did not seem to indicate the selection of one mode of political or cultural practice as opposed to another. There was, instead, simply a new technique: "the substitution of conscious intention for unconscious striving." The new professional class would emerge, assert its strength, and act in the public interest, and Lippmann clearly assumed that the "scientific method" or the scientific approach to problems was by definition in the public interest. Science, he believed, was an impartial force; it was unbiased, unfettered by prejudice or tradition. It offered a method, a way of looking at the world with ruthless honesty and of reaching incontrovertible conclusions.[53]

Even more, it was the instrument with which to forge a new American culture. "Science is the culture under which people can live forward in the midst of complexity, and treat life not as something given but as something to be shaped." It was also "the irreconcilable foe of bogeys. . . . The scientific spirit is the discipline of democracy, the escape from drift, the outlook of a free man." Lippmann was specific about the way this "discipline of democracy" would narrow the range of possible political ideas. "The discipline of science," he wrote, "is the only one which gives any assurance that from the same set of facts men will come approximately to the same conclusion. And as the modern world can be civilized only by the effort of innumerable people we have a right to call science the discipline of democracy. . . . There is no compromise possible between authority and the scientific spirit." Once again, it seems as if "science" was not so much a method of "solving" problems as it was a way of avoiding argument.[54]

Another Progressive Era work probed more deeply into the problem of how American democracy would be reconstructed at a time of massive scientific, economic, and political change. Herbert Croly's *Promise of American Life* was written five years before *Drift and Mastery*, and it shared many of the themes and arguments of the later work. Croly too argued that the changes in America's business system had provided "the opportunity . . . to men of exceptional ability to perform really constructive economic work." He too called for a greater reliance on expertise and an abandonment of the notion that bigness in business was bad. But he introduced an idea that was absent from the works of Lippmann and the techno-utopians and that would prove to be prescient and significant in the ensuing decades: the notion that it was the managed prosperity provided by national organizations that would bring about a simultaneous enhancement of individual and collective life in America.

Americans, he argued, needed to abandon the false opposition they had created in their minds and their politics between the individual and the larger corporate structures of national life, whether those structures were governmental or private. They had to drop the idea that Jeffersonian democracy was incompatible with Hamiltonian concentrations of power. Both sensible leadership by the best and a fulfillment of democratic promise were possible. Prosperity would be the hinge that could join these two together.

"Association," he said most succinctly, "is a condition of individuality."[55] By association he meant effective and placid cooperation between workers and employers and between business and the state. Reasonable state-sponsored economic reforms, for example, would bring rational order to the business realm while increasing "the amount of economic independence enjoyed by the average laborer." It would "diminish his 'class consciousness' by doing away with his class grievances, and intensify his importance to himself as an individual."[56] An associative society that enjoyed a properly "democratic organization" of its political and economic and social institutions would operate "for the joint benefit of individual freedom and development."[57]

Prosperity would do away with class grievances and foster a new kind of individualism. "The highest possible standard of living," Croly argued, was the "essential condition of individual freedom and development." This essential condition was also to be a perpetual goal: "What the wage-earner needs," he wrote, "and what it is to the interest of a democratic state he should obtain, is a constantly higher standard of living."[58]

Croly's book was more realistic than both Lippmann's and Bellamy's about the need for conscious action to achieve the goals he sought. Faith in evolution or scientific expertise was not enough. But while Croly wandered into densely wooded thickets of turgid prose while trying to explain how his new system should be governed, he emerged at the same destination as so many other Progressives: a unified, conflict-free nation that had become "an enlarged individual." The needs of individuals, even when they are enlarged, are easier to satisfy than are the needs of troublesome collections of groups. The society he envisioned was one in which there would not have to be any hard or contentious choices. There would be no winners and losers, only a contented, prosperous, and vaguely Emersonian One, or as Theodore Vail might have put it, One System.

In addition to Lippmann and Croly, many prominent figures urged Americans to shed their antimonopolistic tendencies and to regard the corporation as a benign and even a liberating force in American life. Herbert Hoover argued that monopolies fostered "cooperation" as well as "individuality, equality of opportunity and an enlarged field for initiative."[59] The idea was widespread throughout American thought and letters, and even many liberal reformers argued for the realization of the progressive potentials of the corporation. Thus Lyman Abbott wrote, "Combination both of property and of industry, of capital and labor, is inevitable, because it is the divine order of human development." Frederic Howe believed that the growth of America's corporate economy was evidence that "here as elsewhere evolution has followed a sequence as orderly as it was inevitable."[60] Even in the congressional debates over the Sherman Anti-Trust Act the proponent of the bill, Senator John Sherman, argued that the bill was necessary only to regulate unreasonable restraints of trade; corporate combinations in and of themselves were not to be feared. On the contrary, they "ought to be encouraged and protected as tending to cheapen the cost of production." Corporations, Sherman argued, "are the most useful agencies of modern civilization."[61]

I have examined these writings not out of some belief that they alone created or caused the ideas about science, technology, and politics that would become so prevalent in mid-twentieth-century political thought. Rather, they provided the most vivid example of certain recurring ideas: namely, the notion of an ineluctable process of evolutionary change and the notion of the importance of a new, apolitical, technical, administrative elite in facilitating that change on behalf of the American "corporate commonwealth."

The various writers who expounded on the political effects of science and technology, Bellamy, Lippmann, and Croly in particular, established an intellectual frame of reference to which actual scientists and engineers would adhere with remarkable closeness. They created a climate of thought that was influential but not determinative. "Existing patterns of thought," James Kloppenberg reminds us, "limit possibilities without eliminating new ideas."[62] Scientists and the businesspeople who increasingly relied on them in companies like AT&T depicted the growth of their profession and their companies as part of an "evolutionary" process that was "natural" and apolitical, as natural and apolitical as the vision of social harmony and expert guidance outlined by Bellamy, Carty, and Steinmetz.


Many of the ideas that had been developing from the communities of social scientists, corporate propagandists, and techno-utopianists were given greater institutional force after World War I with the creation of research and advisory groups such as the National Civic Federation, the Commission on Industrial Relations, the Institute for Government Research (the precursor of the Brookings Institution), the Twentieth Century Fund, and the National Bureau of Economic Research. All reflected an elite desire to use the tools of the new social sciences to bring about "wise" reform, to stabilize class relations between employers and workers on some "scientific" basis, and to relocate at least part of the political decision-making process in communities of experts and away from the turbulent democratic masses.[63]

As Olivier Zunz observed, there emerged an "institutional matrix" that consisted of universities, the professions, the foundations, and the government which "reshaped attitudes towards enquiry and patterns of intellectual authority."[64] The very existence of this institutional matrix of public and private research institutions created a new climate of thought and action. It furthered the trend toward the creation of what John Jordan has termed an "engineering model" for American politics and culture.[65]

It is important to realize that the American public in the early twentieth century met the new corporate and managerial elites at least halfway, as evidenced by the popular enthusiasm for the new consumer technologies and new occupational opportunities supplied by the corporate order. Much of the historical literature on elite reforms in the early twentieth century is written as if the public were inert and wholly susceptible to the dictates of the elites. But there was instead a vast working class and middle class of employees and consumers who were using the new corporate economy and its consumer technologies for their own purposes.

The emergence of the radio, the light bulb, the telephone, and the automobile led to a new cultural understanding about the role of technology and the corporations that produced them in American life—an understanding that not only would lead people to adapt to corporations but would likewise compel corporations to adapt themselves to the tastes and the demands of the public.[66] Much of this popular enthusiasm for science and technology overlapped and even reinforced the elite uses of technics and replicated existing social patterns of behavior and prejudice. New machines in the house meant more work for mother, not father. An electrified workplace, indeed, an electrical industry, meant greater corporate power and managerial control. People used new technologies like the automobile and the telephone to reinforce but not necessarily replace older social patterns. David Noble struck close to a central truth about the culture of American technology in the twentieth century when he wrote that "everything changes, yet nothing moves" in a society that was "remarkably dynamic" and yet "goes nowhere."[67]

But these conservative uses of science and technology should not blind us to the innovative and even transforming ways they were also used. Science and technology were not simply agents of repression used by the capitalist classes. They were instead something that Americans used to create their own culture. Americans did not just live with electricity, they lived *through* it. That is, they began to reconstitute their lives in terms of the things that electricity and other consumer technologies made possible such as new kinds of urbanized work as well as new kinds of leisure. These technologies created nighttime office work, domestic conveniences, Times Square and the cosmopolitan nightlife of the "Great White Way," and even a new aesthetic, which one historian has called the "technological sublime." Technology became almost a spiritual medium through which people could express and celebrate the grandeur of human creativity and the exciting new realms of experience and of consciousness that it made possible.[68]

In discussing such notions, we are touching on a positive view of technology that existed in the early decades of the twentieth century and reached its climax in the 1930s. It is de rigueur, to the point of being clichéd, to denigrate this mentality, usually by contrasting the hopeful vision of the 1939–40 New York World's Fair with the horrors of technological war that soon followed. The prewar technophiles came to seem naive and foolish to the post-1945 mind. No doubt the enthusiasm expressed about science and technology by the business sector, scientists, and techno-enthusiasts was exaggerated and excessive. But we must accept that however naive or wrong such attitudes seem to us now, they played an important part in shaping the American discourse about science and technology.

This book is not a history of popular reactions to technological change, but it is crucial to make the point at the beginning of a study of elite uses of science and technology that these elites were operating in a world in which they were not the only ones who had ideas about the meaning of these things, and they knew that. Indeed, elite perceptions of popular expectations and reactions were a constant subtext to the ideas of the scientific leaders of the 1930s, 1940s, and 1950s.[69]

In a political culture in which public opinion and public taste were becoming increasingly important, scientific elites could never simply impose a particular vision on American society. They had to negotiate their way through an institutional landscape and a changing set of cultural values that shaped and limited the way science was used. But intellectual as well as scientific elites were also able to use ideas and institutions to change that institutional and cultural landscape. Carty, Bellamy, Croly, Steinmetz, and Lippmann were not just mouthpieces for particular industries or a particular class. They were trying to assert an important place for themselves in the hierarchy of this new world, and they were doing so as part of a project that they hoped would eliminate the antagonistic class system of the nineteenth century.

Their apolitical strategies, of course, had various political consequences, chief among them being the strengthening of large corporate, national institutions, which were regarded by corporate promoters and many Progressives as the crucial instruments for systematic, evolutionary progress. Making corporations, professions, and expert knowledge so important also had the effect, however, of making *un*important all those people and issues that were outside of this system. Most specifically, the great racial battles of nineteenth-century American politics were muted and covered over by layers of thought and action that simply took it for granted that blacks, nonwhites, the poor of all skin colors, even the religious, were irrelevant.

It is no accident that the 1920s and 1930s were simultaneously the years in which, on the one hand, racial politics reached an intense pitch through immigration quotas, the eugenics vogue, and lynchings, and on the other hand, scientific ideas were used to undermine the intellectual and ideological supports of racism. In a culture increasingly coming under the sway of a corporate-scientific model of social order, the ideas of a Franz Boas would gain greater primacy and overt racism would slowly begin to recede from the American public sphere. But paradoxically, the underlying conditions that made racism possible (or necessary), such as gross economic inequality and cultural, political, and institutional exclusion, would only be ratified by the triumph of this new system. Charles Steinmetz suggested this aspect of corporate scientific thought when he wrote about the need for "racial unity" and the rejection of "racial sectarianism." These were "the first and fundamental requirement of a stable nation." Steinmetz was a man of cosmopolitan and democratic sensibilities, so he could describe this requirement in very appealing terms. He saw racial unity as a benign exercise of assimilation and tolerance. He realized, along with such thinkers as Randolph Bourne, that the great strength of America was the way "native and immigrant assimilate with each other" and each changed the other to create a new, distinctively American identity.[70]

But in the hands of other thinkers and other social leaders, the admirable goal of avoiding racial conflict was translated into the far less admirable strategy of simply ignoring other races and their economic and legal disabilities. By pushing the idea of "race" out of American science, politics, and culture, the new order pushed nonwhite people out of the public sphere as well. This "sinking of the lower classes" and the realignment of the American class system with a new "national class" on the top—the "compromise of the 1930s," as Robert Wiebe has called it—allowed that new national class to control national affairs while local middle classes maintained control over local concerns.[71]

American political culture was being refashioned by scientific, corporate, intellectual, and political elites in a way that excluded the poor and the powerless from meaningful participation in the new national economy and national class system. But to understand how such an exclusionary system was able to survive and thrive, we must also recognize the expansiveness and even the potential flexibility of this new order. Many new political possibilities were available for whites. By redefining the nation itself in a corporate, technological, and scientific model, Progressive Era elites created the possibility for rival interpretations of the meaning of science as well as the meaning of progress and democracy. When those elites shifted the locus of their authority from specific men in specific locales and turned toward a "system" as the instrument for political authority, they simultaneously bought themselves political legitimacy and safety and also made it possible for nonelite groups and even individuals to manipulate that system and shape its character and direction. But corporate economic and political interests and the ideology of corporate science defined the limits of that expansiveness and flexibility.[72]

By the 1930s, something like a new governing consensus had emerged. Society seemed to have been remodeled into what Herbert Hoover called an "industrial machine." In this new technological version of political culture, "the worker, the administrator, and the employer" were "absolutely interdependent on one another," and the purpose of that interdependence was to secure "the maximum of production and a better division of its results" under the elite stewardship of corporate managers and academic and professional experts. Political power in this society became, paradoxically, both more concentrated and insulated at the top and more porous and accessible from below. "The rise of expertise," the historian Thomas Bender has written, "was embedded in a transformation of the public sphere that however imperfectly conceptualized and realized, might well be characterized as democratic in its tendencies and potential."[73]

Alternate visions of a new political economy such as socialism and industrial unionism had been beaten down by the jingoism of the Great War, by the Red Scare of 1919, the Open Shop drive, and the conservative consumerism and ethnic and racial divisions of the public.[74] A corporate scientific creed had triumphed in America, one that stressed the goal of a conflict-free society dominated by large, corporate institutions and administered by experts and professionals who worked in collaboration with those corporate institutions. It was a creed that redefined the promise of American life and democracy in terms of access by whites to consumer technologies and corporate-professional opportunities. And above all, it was a creed that legitimated itself by convincing people that the corporate scientific culture was a natural and beneficent product of "evolution" that would provide a more prosperous, more democratic, and more fulfilling life.

America seemed to affirm the triumph of this new creed by the election of Hoover himself to the presidency in 1928. Even the American Federation of Labor subscribed to it when it proclaimed in 1930 that "trade unions must cooperate actively with management to promote high productivity, elimination of waste, and lower cost of production in organized establishments. Only under these conditions can payment of the highest wages become possible." Hoover could not have put it better: the economic desires of workers could be realized only by accommodating to the ideological prerogatives of a "scientific" corporate capitalism.[75]

But the Great Depression shook the foundations of this new order, and it would reveal sharp divisions that remained in American culture beneath the corporate facade of interdependence and harmony. Neither workers nor their employers would be at peace in a political economy in crisis. Indeed, it is striking how uneasy corporate capitalists remained at the end of the 1920s, even before the full brunt of the depression was felt. Their success in reorienting America's governing ideology around the institutions of corporate capitalism did not eliminate their jittery need to justify their power to a public whose hostility they still feared. Americans had been buying millions of cars and radios and other consumer technologies throughout the 1920s; would they still "buy" the ideology of corporate science during the 1930s? That question and that problem would shape the way corporations and elite scientists used science throughout the 1930s.[76] The economic crisis would also bring conflicts among scientific elites over their role in American political culture, leading some to turn to institutions other than the defensive corporate monopolies in the hope of finding new vehicles for realizing the full scope of their professional and political ambitions.

---

## CHAPTER 2

## SCIENTIFIC POLITICS IN THE 1930S

---

Corporate science expanded and thrived during the 1920s so long as it seemed to be delivering prosperity and progress. But with the onset of the Great Depression the spigot of technological prosperity was shut off. Celebration turned to scorn as critics attacked science and technology for causing the crisis through automation and excessively rapid change. But corporations and their ideology of prosperous social harmony survived the 1930s intact by sticking to their rhetorical guns and disdaining any overtly political role for themselves. The science-based monopolies maintained popular support by convincing the public that they could overcome the slump and provide new technological marvels and more employment. They were able to weave themselves into the cultural fabric of American life in spite of the economic crisis.

Scientists such as Frank Jewett (president of Bell Labs) and Karl Compton (president of MIT) were crucial architects of this strategy. They argued that the depression could be weathered by continued and even increased funding of basic scientific research in the corporations and the universities. This merging of the professional interests of science with the political needs of corporate America would have two important consequences: it would tightly constrict the public political roles of individual corporate scientists and it would lead a faction of scientific leaders, including Compton, Vannevar Bush, Isaiah Bowman, and Ernest Lawrence, to look beyond corporations and toward the state as a source of funding.

These changes were accompanied by intense conflict in the scientific profession. While all scientific leaders in the 1930s agreed on the general proposition that basic research should be used to achieve social progress and harmony, there was still wide divergence of opinion about exactly how that should be carried out, how scientists should act as public figures, and how they should deal with the state. Frank Jewett's career in the 1930s provides a vivid example of these problems, and his story casts a revealing light on how the scientific politics of that decade were also changing America's political culture with regard to the relations between the public, experts, and large corporations.

Jewett represented a vision of scientific expertise and professionalism that was seamlessly woven into the ideological interests of the corporation for which he worked. Corporate monopolies were essential for fostering the kind of corporate science he had chosen to practice. That choice had not been an easy one when he made it at the turn of the century. Both Jewett and AT&T embraced corporate science with some uneasiness, but both thrived as a result of their decision. AT&T was wary of the costs; Jewett was wary of the low prestige in which his new occupation was held. After earning his Ph.D. under A. A. Michelson at the University of Chicago in 1902, Jewett taught for two years at MIT and then accepted an offer to join Western Electric's research and engineering department in Boston at an annual salary of $1,600 per year. Michelson strongly objected to Jewett's choice. He felt the younger man was "prostituting his talent" as an applied scientist. Jewett would be instrumental in improving the scientific quality of AT&T's research, but nonetheless, he still felt the sting of Michelson's criticism thirty years later. In a 1938 speech reciting some of the history of corporate science, Jewett observed, "Although he relented somewhat in time, I think I never regained the respect of my old chief, Michelson, after I forsook the halls of learning for the walls of industry."[1] This idea of justifying "applied research" or of showing how basic and applied research were inextricably linked would form a crucial part of the professional identity of the corporate scientist in the 1930s. As we shall see, it also emerged as an important aspect of the self-image of scientists who worked on military projects in the postwar era.

Jewett continually argued that the professional experience of the corporate scientist was not fundamentally different, in an intellectual sense, from that of the academic, or "pure researcher." Industrial research was always a mixture of "problems of current operation" with work of a "fundamental character." There were no doubt some, wrote Jewett, probably thinking of his mentor's rebuke, "who profess to see something akin to prostitution of scientific research in the thought of employing it for anything save the acquisition of knowledge for its own sake, [but] their number must be small." The distinction between basic and applied science was becoming blurred to such an extent that it was meaningless. The work on the transcontinental project illustrated this point perfectly: it required fundamental knowledge about "the production of electrons and the conduction of electricity through gases of high and low refraction," while it also demanded purely technical skill in implementing these ideas in the telephone apparatus. Certainly, Jewett conceded, there were differences between an academic researcher and a corporate scientist; but the differences concerned motive only, not scientific sophistication.[2]

Jewett and his colleagues were practicing science in a new way. They had to formulate "technological theories" to fully understand the scientific principles at work in their technical creations. There was no rigid line between mechanical tinkering and pure theoretical exercises. If anything, as the Federal Communications Commission (FCC) noted in one of its reports about the company in the 1930s, fundamental knowledge was the by-product of the company's efforts to create "practical apparatus." Theory derived from practice, not the other way around.[3]

The professional identity Jewett reflected in his writings was not embodied in a fully worked out structure of schools and traditions; there was no formal process of becoming a "corporate scientist," but it was a real and distinctive professional identity nevertheless. Scientists who joined corporate enterprises still trained under the same curriculum as their academic colleagues. But the growing institutional connections between the corporate labs and the major technical schools like MIT and Cal Tech, and even in some of the elite, nontech schools such as Harvard, provided new professional opportunities for young scientists.

The terms of this new professional identity were filtering down to young graduate students. James Conant later recalled that among the thirty-three students awarded Ph.D.'s in chemistry at Harvard between 1907 and 1917, "there was hardly one who did not intend to become a college professor." Ten years later, "the majority of those who received the advanced degree would enter industrial employment either at once or after a few years."[4]

Opportunities for corporate research were just beginning to open up when Conant was an undergraduate. He recalled a lecture by Willis Whitney, in which the first director of the research laboratory of the General Electric Company talked about a young scientist who had been induced to leave his academic post to work in his lab. He had offered the professor "an opportunity to carry on in the GE laboratory the same line of research he had been pursuing," and he was confident that "once the new recruit had a chance to look around, he would find more fascinating questions in industry than the problem he was bringing with him." At length, "the erstwhile professor[5] lost interest in what had once been his sole concern. The main burden of Whitney's remarks that evening was to the effect that industry abounded in problems which lent themselves to attack by a trained research man irrespective of the field in which he had gained his experience. The story made a deep impression on me."[6]

By the 1930s, enough institutional links existed among the research laboratories and their directors to suggest the formal structure of a profession. Part of this was sheer increase in size. In 1890 there were only four industrial research labs in America. In 1930 there were one thousand.[7] Leading corporate scientists started to provide institutional structures to strengthen and advance their nascent profession. Charles Reese of Du Pont created an organization in 1923 called the Directors of Industrial Research. It was initially formed under the auspices of the National Research Council (NRC), but the group proved so useful to its members that it broke away from the NRC and became a self-sustaining entity. Its purpose was to allow the lab directors like Reese, Jewett, Whitney, and others to meet monthly and discuss common problems and strategies. The members of the group periodically hosted tours of their labs for their colleagues' benefit.[8]

The experience of the industrial research lab showed that scientists could work effectively with businessmen, and, perhaps more remarkable, even scientists from different fields could cooperate with each other.[9] The general practices of corporate science made it easier than in academia for scientists of different fields to collaborate and move across intellectual boundaries. Laboratories run by corporations or by foundations like the Carnegie Institution in Washington became places where scientists could engage in work on the "borderland" of different disciplines. Such research would be duplicated in academia only with the creation of Ernest Lawrence's Radiation Laboratory at Berkeley. One student of the chemical sciences in this period concluded that "America's academic and industrial cultures had moved far toward integration,"[10] and this integration "had not fettered these scientists, but had freed them."[11]

The appeal of corporate science lay not just in the new scientific problems or the combinations of practical and theoretical problems it presented the researcher. It also offered a new social role for scientists who now had the opportunity to rise up the ranks of the nation's largest corporate organizations. Jewett emphasized this aspect of his professional culture and repeatedly stressed the integral role of corporate scientists to large business enterprise.

In any successful corporate research lab, he wrote, the "head of the research organization participates directly in the consideration of matters affecting company policy." The most successful science-based businesses were those in which the "research function" was "an integral part of executive management."[12] The leader of the lab, he stressed, "should be a man with a broad outlook, a full appreciation of all of the factors of the business problem, and one who can sympathize with the varying points of view which he will encounter and which he must harmonize if the resultant activity is to be that of a smooth-working machine." The rank and file of the organization should be "competent young men" who could be groomed for long careers in the company. Older managers should "provide them with the facilities and encouragements for growth, and ultimately . . . make leaders of them."[13]

Jewett envisioned a future in which engineers would play an increasingly significant managerial role in the science-based industries. "We must look in increasingly large measure," he said in 1924, "to the group of technical graduates for the executives of the future. The problems of the industries which executives will be called upon to administer . . . are becoming increasingly involved in the just appreciation of fundamental physical science. . . . From the group of well-trained engineers should come many great executives."[14]

Jewett's conception of his professional and public identity depended on the assumption that the interests of corporate science and corporations were identical. But were they? Scientists gained popular acclaim and media attention only to the extent that they appeared to be "wizards," bringing on new technological marvels and enhancing the comfort and luxury of the individual American. In a culture in which the public expected scientists merely to be wizards, there was little cultural space to argue, as Jewett would in the 1930s, for scientists to engage in the wider application of the "scientific method" in America's political affairs. Corporate scientists may have thought of themselves as significant partners, but that was not necessarily how the public or corporate executives saw their role.[15]

Carty, in the 1920s, had played to the "wizard" theme. Jewett, in contrast, while in no way diminishing the significance of the accomplishments of the scientists, placed much greater emphasis on their methods. It was the scientific method, Jewett argued, and its application in business that made these vast material achievements so popular and so significant. By emphasizing the scientific method, Jewett implicitly argued for the social significance of the scientist as a public expert, independent of his affiliation to any corporate organization.

Jewett believed there was something unique about the talents of the scientist that justified both his institutional importance and his public usefulness. But his vision of the role of scientific expertise in America was becoming too elitist and didactic for it to be palatable to an increasingly assertive public. In 1929, for example, in a speech at his alma mater, Cal Tech, he told the assembled engineering graduates that they "were destined to play an increasing part" in the "political machinery" of the country. This was all to the good, he explained, because college graduates, and science and engineering graduates in particular, constituted the country's "ruling class."[16] Such notions, Jewett conceded, flew in the face of conventional American shibboleths about "the people" ruling. But whether or not it was acknowledged, "there is a ruling class—not, to be sure, a ruling class defined by right of birth, but one to which any may aspire to belong." Science and technology students had a special place in this elite because they, in comparison to their colleagues in the humanities, understood that "there is such a thing in the world as absolute truth." They knew there were "inexorable laws whose consequences cannot be avoided by any plea of ignorance or by any amounts of sophistry or mental gymnastics." The technical student had developed the amount of skepticism necessary to check his results and assumptions, but he was ultimately respectful to the existing order. He had "learned to be that anomalous thing—a youthful conservative about some matters."[17]

Jewett hoped that the "social and political machine" would recognize the potential power of expertise and turn "toward those who can guide it with real understanding." By no means did he want scientists to assume power. "I for one would hate mightily to live in a society turned over largely to scientists and engineers. On the other hand I desire very much to see their influence more directly potent in developing the social and political controls for the things of their creation." He wanted to see more scientific and engineering knowledge "on the bridge of the ship of state and not wholly confined to the engine room."[18]

For corporations like AT&T, however, even such an abstract notion of expertise jarred the supreme tranquillity and quiescence the company wanted to convey regarding its public role. AT&T simply wanted to be left alone by the government and the antimonopolists and allowed to make its millions in peace. It did not want to endorse the idea that scientists, especially its own scientists, could launch themselves into public debates and take controversial positions on volatile political questions.

AT&T wanted to use science to help fashion a public image of itself as a benign, progressive institution that provided a useful public service. Consequently, publicity efforts in the 1930s depicted the company's technical expansion as "natural" and even patriotic. A 1936 pamphlet, titled *The World behind Your Telephone*, illustrates the gist of its publicity rhetoric. A company publicist wrote that Bell's invention "included an ideal whose creative powers have built a new and expanding world—a science, an organization, an industry, and a system. And as these have grown, the outer world they serve has been brought more and more within the confines of one neighborhood." If that was not comforting enough, the reader was assured that "the Bell System is organized on a plan much like that of the Federal and State governments," with the twenty-four regional companies playing the role of the state governments; AT&T, the parent company, was, of course, the central government. A few pages later the analogies reached their hyperbolic climax: the physical plant of the company is so massive its wires could wrap themselves around the earth and moon 168 times, its service so far-flung, its employees so numerous, so well cared for, that "the modern telephone company serving an extensive territory is a world in itself."[19]

The company's brass wanted to stamp out any notion that the Bell monopoly might have political interests. In 1941 AT&T published a book under the name of one of its vice-presidents, Arthur W. Page, titled *The Bell Telephone System*, whose central purpose was to rebut a critical FCC report issued in 1939. Page took special pains to refute the idea that the company was given to throwing its weight (i.e., its money) around state legislatures that had jurisdiction over rates. "The Bell System," Page wrote, "has no political influence and wants none." Furthermore, company policy forbade employees from participating in a political controversy while identifying themselves as AT&T employees.[20]

Far from being a sinister, manipulative monopoly, the company was an exemplary institution: the Bell System, Page wrote, "mixes reason with authority and routines with initiative so that more than a quarter of a million people can engage in a complicated art, each using his or her brains, and yet have the whole thing move toward a desired end without confusion." The organization that permitted this mixture of individual action and group purpose was "as significant as the discovery of new facts about nature that allow what is called technological progress." Echoes of Croly as well as the techno-utopianists reverberated in Page's prose. The science-based corporate organization was presented as a thing "discovered" in "nature," which had solved what one might call the other American Dilemma: it had preserved democratic individualism in a collective, bureaucratic political economy.[21]

Not surprisingly, Page also invoked the master paradigm of evolution: just as "the human body is supposed to change every seven years," the "telephone plant is renewed by a similar evolutionary process." Page's words, like the company's other publicity material, offered an image of its technology that was devoid of human control. The "System" was an awesome presence that mystically unfurled itself from the coils and cables of Alexander Graham Bell's invention. The public-spirited managers of this system merely kept it expanding on its "natural" and benign path, rather like the administrators of Bellamy's utopia. This was not a view of technology and culture that implied an assertive public role for scientists and engineers.[22]

AT&T was not the only company that employed such rhetoric. Other companies tapped into this cultural vein as well. GE also stressed the unifying and even utopian aspects of its scientifically created products. Its publicity writers drew upon Bellamyesque rhetoric that depicted society as a machine beneficently tended by the company's engineers. Such language enabled engineers to advance scientific and engineering ideas in place of the traditional concepts of politics. Social life and social change were the result of wisely managed engineering, not political conflict.[23]

Other leaders of major industrial research organizations also shared Jewett's notions of corporate science as something that embodied the evolutionary process. Charles Kettering of General Motors (GM) noted that the new technological culture in America was to be one of "change, change, change, all the time; and it is always going to be that way." There would be no going back to the older world either, as Willis Whitney noted. "There are processes," he wrote, "like the scrambling of an egg, that are not reversible, and such are the mighty processes through which our industrial civilization has been evolved."[24]

Paradoxically, the corporate men using this rhetoric of ceaseless, irreversible, and even revolutionary change were all rock-solid conservative Republicans. The paradox unravels, however, when one realizes that they deployed this rhetoric as an argument against political radicalism that might threaten corporate interests. But what these men did not seem to realize was that if political responses to scientific, technological, and corporate change were to be pushed out of public life in place of a discourse of benign, techno-utopianism, then *all* political language, conservative as well as radical, was out of bounds. Jewett is the sharpest example of this. His political utterances in the 1930s implied the sort of abrasive, political position that the corporate propaganda of companies like AT&T and GE was trying to transcend. He seemed to think it was possible for scientists and engineers to have public influence by acting through powerful institutions. But Jewett's own company made it clear that the accomplishment of individuals was less important than the allure of a positive and abstract corporate image.

The distance the company sought to keep between the lab's scientists and the public was suggested to Jewett by one of his colleagues, a vice-president in charge of publicity, John Mills, who explained that "the Bell System has all the usual motives for advertising for external good will. None of these applies directly, however, to the Laboratories as an individual organization, although certainly a well known laboratory would be an advantage to the Bell System in wider public acceptance and greater friendship." There was, Mills concluded, "little reason for the Laboratories itself to advertise although excellent reasons for the Bell System to advertise the Laboratories." The company's public relations, in short, would not stress the company's scientists as a distinctive group.[25]

General Electric was also loath to use its scientists for advertising purposes during the 1930s. When a manager in the refrigerator department suggested to President Gerard Swope that the industrial research scientists be used in an advertising gimmick to apply a scientific seal of approval to the company's refrigerator, Swope rejected the idea. It was demeaning to the scientists—and besides it was bad advertising since scientists had lost their luster in the public mind in the early years of the Great Depression.[26]

In Jewett's case at AT&T, the differences between his and the company's conception of the scientist's role suggest an important political difference and not just a marketing issue. Jewett, unlike his employer's publicists, regarded science as a potential problem as well as a boon. His deeply conservative mind was uneasy over the disruptions wrought by his work. "I hate revolution," he once told an interviewer, but he recognized the revolutionary character of scientific innovation, and it disturbed him. "There is nothing that has sprung from the workings of the human intellect during all recorded time," he wrote, that could compare "in power and awesome grandeur with the change which came into being about the end of the eighteenth century . . . the so-called scientific method." It wrought rapid "cataclysmic changes" that had been "catastrophic" in their effect on "age-old customs of life and long established methods of social control and social intercourse."[27]

He did not adopt the view of his employer and other corporate promoters that all that flowed from the scientific laboratory was benign. "Contrary to much common belief," he wrote, "science is not uniformly beneficent but, like many animals, is prone to destroy its own offspring. . . . The unknown man in the laboratory with his test tube and chemicals or his microscope and galvanometer can be far more destructive of our hopes and aspirations than all the known forces marshalled against us."[28]

The potential destructiveness of science, he believed, compelled scientists to play a significant part in shaping its social application. Part of that responsibility involved educating the public. As he argued in 1932, "We must find a way to a truer and widespread indoctrination of [science's] possibilities and limitations." He never expected the general public to develop a fluency in the esoteric and demanding disciplines of the basic sciences, but he believed "a real and not a superficial understanding about a few underlying things of science [was needed] on the part of those who are compelled to make great decisions concerning future happenings."[29]

This did not mean, however, that scientists had to exercise direct political or social power over the implementation of technology and science even though, somewhat contradictorily, his views suggested that scientists were better suited for such leadership roles. "The things of science," he argued, "can be left to take care of themselves." But it was particularly important for Jewett to achieve some universal consensus on the social implications of science and technology. Only if there were "some conscious group point of view in these matters can society avoid drifting like a rudderless ship, buffeted by every wind that blows and with each gale more violent than its predecessor." Jewett had no use for the radical technocratic program of bestowing ultimate political authority on society's engineers. He wanted the technical elite to exert its sway through the corporate system, not apart from it.[30]

Jewett devoted considerable energy to several cultural efforts in the 1930s, all of which were designed to "educate" the public about the role of science and scientists in the process of technological and social change. Beginning in 1928, he was chairman of an advisory council that had been formed by the NRC to assist in the preparation of the Chicago World's Fair in 1933. From start to finish the Chicago Century of Progress Exhibition was a forum for corporate promotion and propaganda. One of the guiding spirits of the expo, Rufus Davies, a Chicago oil man, had enlisted the help of the NRC to get that elite group's imprimatur in promoting the essential idea that American prosperity depended on "a close alliance between men of science and men of capital."[31]

In planning the exhibits for the fair, Jewett hoped to carry out a "quiet unconscious schooling to the thoughtful people" who strolled through the displays. Visitors would "pass through the fair and go out of it, largely without any consciousness of having been educated." But they would have learned that "science is at the root of most of the material things and many of the social things which make up modern life." The fair would contain a large Hall of Science, along with a separate hall for displays about applied research.[32] Jewett would also be involved in the famous 1939 New York World's Fair, where the emphasis on corporate promotion drowned out any effort at transmitting "scientific knowledge." The grandiose corporate exhibits that promised a utopian "world of tomorrow" served to stir the wrath of many scientists who felt that their profession was being buried in the Barnumesque tactics of the businessmen. One student of these fairs has concluded that the "exposition designers hoped to adjust Americans to a political culture increasingly dominated by corporations."[33]

But Jewett's role in such efforts cannot be reduced to a simple assertion of corporate promotion. He wanted both to educate and to propagandize, and the important thing is that he could not see the difference between the two goals. This failure is best illustrated in his most extensive experience with the cultural treatment of science: as a trustee and president of Henry Towne's museum, now renamed under its new incarnation as the New York Museum of Science and Industry. The conflicts and changes in that institution during Jewett's involvement from 1925 to 1942 offer the clearest example of how out of touch he was with the prevailing corporate manner of using science and scientists.

By 1936, when Jewett took over the presidency of the museum, the directors consisted entirely of other corporate executives. The museum's policy was geared almost exclusively toward serving the promotional and advertising needs of business. But in spite of these changes, the museum still lacked an endowment, and none was forthcoming. The proliferation of radios, mass circulation magazines, and other popular mass media left companies little reason to spend much time or money on a static, old-fashioned advertising medium such as a museum. Jewett failed to understand this new economic reality. He was genuinely perplexed by the indifference of potential corporate contributors, an indifference that seemed to increase with the financial difficulties of the museum. In 1941, he was unable to get even $150,000 to cover the museum's operating expenses for that year. "I cannot believe," he wrote a museum colleague, "that American industry is shortsighted enough to allow a thing like the Museum, with a long history of achievement, to pass out of existence for a paltry $150,000."[34]

For Jewett, the education offered by science museums had profound political significance. He hoped that with "every bit of new scientific knowledge acquired" there would be "an addition to the strength of the social structure, and not a revolutionary threat to the social order." How this was supposed to work out in the real world, he explained only vaguely. He offered some insight in a 1934 speech regarding the role of science museums. "We . . . need," he said, "to have some group understanding of the fundamental laws of, say, modern transportation machinery and of the further possibilities and limitations of science which underlies it, if we are to set up satisfactory rules governing transportation in the national interest."[35]

But in that very same 1934 speech he also went on to say that the science museum could be a place where the public could go to look at mock coal mines and "understand" that "it would be foolish to yell for working conditions which required galleys twenty feet high in all mines no matter how desirable they might be otherwise. We would know that nature forgot to provide them." A museum that displayed new industrial machines alongside older models would allow visitors to determine for themselves whether the "sob stuff" about the stretch-out system and the exploitation of industrial workers was true. This was not "education," it was propaganda.[36]

He could say with regard to the public role of scientists that "part of our job along with being good artisans is to be good teachers. And this teaching cannot be done merely by putting on a good vaudeville show even if it is interesting and makes people open their eyes and their mouths. We should try to make them understand the real values back of the glitter." Jewett regarded such museum exhibitions as "one of our great educational tools," and he felt it was essential that "the mass of us must come to some knowledge of certain fundamentals of science and the scientific relationships which control much of the world we live in, if as a group, we are to avoid chasing forever those perpetual motion phantoms which we have chased so often in the past."[37]

But his inability to distinguish between education and propaganda was indicative of the contradictory nature of his role as a corporate scientist. The use of anachronistic terms like "artisans" is revealing. At least a part of Jewett's mind still saw the public role of the scientist in America in the 1930s as comparable to that of a paternalistic master craftsman of the pre-industrial era. He envisioned a cultural context in which ideas and values could be transmitted directly, the way a master would "educate" an apprentice in a shop. Jewett's interest in science museums is also indicative of this simple view of cultural transmission. The museums, he believed, were places where members of the working classes could come and be "educated" by the professional and managerial class. But these static, almost quaint notions about the transmission of politics and culture were being superseded by a corporate system that presented itself as a seamless web of progress and prosperity in which there were supposedly no class distinctions. The transmission of cultural and political values in this new system was scattershot, superficial, and increasingly relegated to the media of advertising and public relations. Corporate monopolies wanted only to blunt class conflict and to transmit a benign image of harmony and progress. It was irrelevant whether the public "understood" the scientific practices taking place in the research lab. It mattered only that the public not get angry enough to support the reformers who periodically mounted the antimonopoly bandwagon.

Not only were Jewett's ideas about the pedagogical role of scientists becoming anachronistic but his brand of conservatism was becoming a lifeless relic. The emerging liberal concern for the weak had little place in his thinking. Jewett had little sympathy for the "under-privileged and the problems which their existence present," believing that "we will always have them with us, at least until some slow process of biological evolution may result in all human animals being alike."[38]

But his cultural activities suggest another side, a paternalistic sense of social responsibility for the "revolutionary" new technologies he was helping to introduce into the world. His sense of responsibility was narrow, to be sure. But it bespoke a political sensibility that was being overwhelmed by the corporate, bureaucratic interests that Jewett and his colleagues were strengthening. In Jewett's mind, scientists needed to transmit their ideas to the lower classes to ensure that those ideas stabilized, rather than threatened, the social order. Such notions of class stewardship had little relevance in a political culture in which the very notion of classes was being dropped in favor of an ideology that depicted all Americans as consumers who luxuriated in the benefits of corporate productivity.

There was, however, a younger generation of corporate scientists emerging onto the national scene in the 1930s. They were enthusiastic proponents of the corporate-scientific creed, but unlike Jewett, they would eagerly turn toward the state, which they hoped would join the associational order that was already taking shape among corporations, the professions, and the universities.

The kind of partnership envisioned by scientists such as Karl Compton, Isaiah Bowman, and, later in the 1930s, Vannevar Bush and James Conant, represented a very different conception of the public role of corporate scientists. This younger generation shared many of Jewett's conservative political values. Compton and Bush, for example, were, like Jewett, no friends of the New Deal. But they shared a very different professional and organizational philosophy.


Scientists such as Karl Compton, the president of MIT, and Isaiah Bowman, president of the National Research Council and later of Johns Hopkins, had a vision of the public role of scientists that was notably at odds with the ideas of men like Jewett and Robert Millikan. For the younger scientific leaders, it was less important for scientists to "educate" the public than it was to establish institutional connections with the state so that scientific research could continue unhindered during the depression.

Throughout the country's history there had always been a connection between scientists and the federal government.[39] The group research or "big science" model adopted by corporate science in the 1920s originated in government-sponsored research projects in the nineteenth century. The creation of the National Research Council in 1918 established a weak link between the profession and the state. But scientific leaders in the 1920s were not yet ready to admit a large or permanent role for government in the funding and practice of science.

George Ellery Hale's failed effort at creating a national endowment for basic research was illustrative of this problem. Like Jewett, Hale wanted science and scientists to have a greater influence on shaping national life, but he wanted that role developed through the private rather than the public sector. Hale, the director of the Mount Wilson Observatory, was instrumental in creating the National Research Council, which he convinced Newton Baker and Woodrow Wilson to charter as a subsidiary of the National Academy of Sciences (NAS) in 1918. He shared with engineers like Carty the idea of science as an integral force in American culture, and he hoped the profession, and the National Academy in particular, would grow in "standing and prestige" so that it might "influence and help the government" and treat with it as an equal national institution.

Hale hoped that Robert Millikan would become the dynamic leader that such a potent and prestigious institution would require. But Millikan did not share Hale's faith in the academy as a vehicle for enhancing the prominence and usefulness of science in American culture.[40] During the mid-1920s Hale tried to organize a national endowment for pure research to be funded by the large science-based businesses. He hoped such an enterprise would cultivate alliances between "men of science" and "eminent men of affairs." Scientists should especially "enlist the active cooperation of men who have established reputations in administration and finance," he wrote, so that through such associations they might enjoy greater confidence in their own abilities to "deal with large administrative and financial projects."[41]

The members of the board of Hale's endowment would have included John Carty, Charles Evans Hughes, Andrew Mellon, John Davis, Owen Young, Felix Warburg, and Julius Rosenwald. The chairman of the endowment was to be Herbert Hoover. The effort failed because only a few monopolistic corporations were willing to contribute to research that might have benefited some market rival. AT&T, as an unrivaled monopoly, had few qualms about contributing to the research fund (Jewett, in fact, was an active member in the effort). But most businesses had little reason to subscribe.[42]

Hale's agenda of achieving greater prominence and influence for scientists was continued and extended by such men as Isaiah Bowman, Karl Compton, and later, under wartime conditions, Vannevar Bush. This younger generation of scientific leaders recognized that the power and wealth of the federal government and not just the doubtful generosity of business would have to be relied on to fund scientific research properly. But scientific leaders in the 1930s disagreed sharply over this question of government funding of science. Jewett was one of the most passionate opponents of increasing the state's presence in the profession. He found himself on the losing side of this debate with his professional colleagues, several of whom he helped promote to positions of influence. He was involved, for example, in the search committees that led to the appointment of Bowman as president of the NRC and, later, Johns Hopkins; Karl Compton's selection for the presidency of MIT and Bush's for the presidency of the Carnegie Institution in Washington were also owing in part to Jewett's intervention.

Bowman was more in tune with Hale's vision of the public role of American science, and he was willing to have scientists work through government institutions to achieve greater prominence. He wanted the profession to produce leaders who were capable of working with the government in addressing the nation's problems. When he cast his eyes over the membership of the National Academy of Sciences in the 1930s, however, he was not encouraged by what he saw. "An alarming percentage," he wrote, "consists of men of technical or professional accomplishment who have no sense of organization, no sense of responsibility for science in government, no breadth of outlook with respect to relations and obligations of science to society."[43]

He wanted to make up for all of those deficiencies, and he lobbied Harold Ickes to get a Scientific Advisory Board (SAB) added to the three advisory panels that already constituted the Industrial Recovery Board. Bowman succeeded, but his success was limited. The SAB, which included Jewett and Millikan as well as Bowman and Karl Compton, had no ability to gain access to government money and it had only a two-year life. There were internal problems as well. W. W. Campbell, the president of the academy, resented the placement of the SAB under the aegis of the NRC, rather than the older and more prestigious National Academy.[44]

Karl Compton, who was also eager to tap into federal funds for research, proposed a $16 million appropriation for the years 1933–39 for a public works program for science. That proposal was rejected. The next year, in 1934, he advanced an even more ambitious proposal that would have allocated $75 million to the SAB over five years for scientific and engineering research outside of the government. That too never materialized.[45]

Compton and Bowman admitted temporary defeat. The few members of the cabinet they had counted on, and indeed the president himself, could not or would not support their initiatives. Jewett was not terribly disappointed over the failure of these efforts to shake money from the federal tree. He wrote Compton on December 6, 1934, to express his fears of greater government intervention in American science. He observed that "appropriation of Federal funds is a very grave question which goes to the root of the whole matter of state participation in and control of functions which hitherto we have jealously guarded as the affairs of individual or non-political cooperative effort. Do we of the Science Advisory Board and the National Academy of Sciences wish to go on record as advocating so revolutionary a departure from our past philosophy?"[46] John C. Merriam, the president of the Carnegie Institution in Washington, was also skeptical and even a little fearful of Compton's proposals. "It is not at all clear," Merriam commented in a November 1934 memo, "that the spending of large sums of money is the thing most essential or of first importance in furthering science at the present time." He saw "grave danger" in the profession relying on such funding "from an organization under political control."[47]

Karl Compton thought scientists had no choice but to seek federal funds. On December 9, 1934, he told his fellow SAB panelists that money for scientific research could come only from "industry, philanthropy, and government. We know that it would be fruitless to turn to either of the first two of these alternatives."[48] Compton wanted to place the scientists on record in support of some effort to alleviate the distress. He not only wanted to repair the damaged public reputation of his profession, but he wanted to mount an ideological counterattack and show that far from being a threat, scientists were a potential salvation. "We technologists," he wrote, "wondered if we had any right to be alive, let alone to do our job, as we heard and read the theories of those who believed that higher standards of living are assumed by curbing production, not stimulating it."[49] In a proposal titled "Legislation for Reduction and Relief of Unemployment," the MIT president wrote: "The public can do either of two things: (1) It can let things drift . . . and meet emergencies inefficiently without plan or organization," or it could put into effect "certain well considered measures which will tend to prevent unemployment" and lead to the evolution of a "stable, smoothly running social order."[50]

The ideological framework within which Compton operated is important to note. He was trying to place the scientific profession in a position of national prominence and significance; he wanted his profession to be more than just a repository for laissez-faire, big business enthusiasts. He was clearly not an enemy of the corporate business community, but like Jewett he believed that scientists had a role to play as public critics and as conservators. Scientists were obligated to point out the waste and inefficiency of business and not just act as full-throated boosters.

There was a noticeable conservationist streak in Compton's ideas about the role of the SAB. Such bodies were necessary, he believed, because unbridled business expansion and the consequent exploitation of resources had sapped America's natural wealth. Prosperity could no longer be expected to flow from the plundering of nature. The genius of scientists would have to be applied to fashion new innovations in a world of scarcity. In his proposal to Ickes and Franklin D. Roosevelt—"Put Science to Work"—Compton argued that scientists could help develop new synthetic fuels; they might be able to develop silk from wood or rubber from weeds. Just as corporations like AT&T and GE had reoriented themselves through the exploitation of scientific research and expertise, Compton seemed to be arguing, so too should the government and the polity.[51]

But could the sheer "evolution" of science and industry be counted on to pull the country out of its economic depression? Or did the crisis and the "evolution" of science and industry itself require a more public, active, and critical role on the part of scientists in adjusting their accomplishments to the existing social order? Scientific leaders of the 1930s had come up against the same problem posed by scientists and Progressives a generation earlier: did scientists have to participate directly in government to manage the social effects of their creations? Or could they keep at arm's length from the government and let the private sector handle the process of technical and scientific change?

Scientists such as Jewett and Karl Compton seemed to endorse both perspectives without perceiving any contradiction. Jewett, as we have seen, saw dangers as well as benefits from scientific innovation. Scientists had to be involved somehow in controlling or eliminating those dangers. Compton thought the "somehow" lay in simply getting the government to give scientists money to promote more basic research. His efforts regarding the failed Scientific Advisory Board and some of his experiences at MIT revealed how poorly he understood the ideological and professional implications of his commitments.

He wanted the government to supply $75 million over five years, but he would not say how it would be spent or who would get it. From Compton's perspective, this insistence was not unreasonable. Had not corporate scientists been given considerable discretion in using their companies' money for research projects that might not, on their face, have appeared to be conventional or immediately practical? The experience of corporate science showed that the fruits of research could not always be predicted beforehand

and that a certain degree of faith or risk was required in allotting money to particular individuals or particular projects.

Harold Ickes, however, opposed giving such a "large 'free fund' " to scientists without any supervisory check. He decided to bring the SAB under the administrative aegis of the National Resources Board, directed by FDR's uncle Frederick Delano, who considered Compton's requested appropriation of $75 million a "staggering" amount. Harry Hopkins, who agreed with Ickes about very little, shared his bureaucratic rival's uneasiness about the SAB. Hopkins convinced FDR to mandate that 90 percent of any research project funds allocated from the Works Progress Administration be devoted to employing people on the relief rolls.[52]

Compton's efforts to gain federal funding should not lead to the mistaken conclusion that he was a supporter of Roosevelt's policies. The New Deal, he wrote, "has been played with the same old greasy cards. It has consisted of taking the aces and jokers from one group of players and handing them to another group." But with investment in research, Compton continued, "a new deck can be created with a lot more aces. . . . Research creates wealth, does not merely reshuffle it."[53]

This antagonism did not go unnoticed. Compton's arguments about pursuing abundance through scientific innovation ran directly in the face of the Roosevelt administration's early New Deal approach of cutting back on production in certain industries. Even some allies in the administration began to turn against the scientists. Henry Wallace, who shared many of the scientists' beliefs about the likely benefits of increased investment in research, attacked conservative scientists as "stand-patters" who were not willing to do enough to help cure the depression. Wallace wanted more attention paid to the social as well as the natural sciences. Roosevelt himself got into the act a few years after the SAB episode. In the fall of 1936, he wrote Compton a letter that was published in *Science*, along with Karl Compton's response, under the title "The Responsibility of the Engineer," in which he argued that engineers had to "also consider social processes and problems . . . and must cooperate in designing accommodating mechanisms to absorb the shocks of the impact of science." The president asked engineering educators to consider whether their curricula were balanced to provide students with "the vision and flexible technical capacity necessary to meet the full range of engineering responsibility."[54]

Karl Compton replied in fury. "I cannot but wonder why your exhortation has been directed specifically towards engineers, for surely we would agree that similar breadth of knowledge and training is also urgently desirable among business leaders, economists, and politicians." He decried the tendency "in some quarters to make science the major scapegoat of our social ills, from which social planners will rescue us." He then went on to restate the "more-science-means-more-employment" line and recited the history of the SAB, emphasizing the administration's complete indifference to the board's suggestions.[55]

Compton was right about FDR's indifference to the ideas of the SAB members. The president thought very little, and then only superficially, about the role of science and technology. But Compton himself did not understand precisely what relationship he wanted to create between the scientific profession and the state. In 1932 he had himself endorsed the idea of planning. "Intelligent people," he said, "are coming more and more to the conclusion that the real salvation of our civilization will have to come through the adoption of intelligent scientific planning." An economic policy guided by the spirit of letting "nature take her course," he warned, "leads inevitably to chaos." How this "scientific planning" was supposed to come about without some guidance or control from the government was a question that Compton seemed unable to answer clearly or even to formulate.[56]

The ambivalent and even contradictory attitudes of the profession's leaders can be vividly seen in a colloquy that was transcribed at an SAB meeting in September 1934. Participants discussed the proposal broached by a member of Congress to create a New Industries Board. Vannevar Bush had been delegated by the SAB to draw up a response to this idea, and his report and the general reaction to the idea among the SAB members revealed how weakly they were inclined to follow through on their own rhetoric about the creation of new jobs. "I have seen enough cases of this new industry stuff," Jewett groused, "to realize that for every really good thing that may crop up there are a hundred things that are no good, and the problem of picking the grain of wheat from the bushel of chaff is terrible."[57]

Several other members of the board echoed Jewett's skepticism. Simon Flexner urged caution before "entering upon so revolutionary a step" as getting the government involved in funding scientific research, especially in the universities. John Merriam noted that "no means have been found for aiding without exerting control over the activities" being funded.[58] Karl Compton weakly acknowledged those concerns but did not really address them. The "entry of government into scientific research," he agreed, "might do far more harm than good, and we can see ways, too, in which it would be definitely advantageous." Jewett summed up the prevailing sentiment most vividly when he said, "The matter merits our very careful consideration, but it is a thing that is loaded with dynamite."[59]

When Bush presented his report on this subject he challenged the need for a New Industries Board. What was really needed, he argued, was patent reform. Bush, however, also adopted an argument that did not seem to resolve the political concerns of the older leaders. He favored government support for scientific work in educational institutions and science museums: "The Federal government can be and should be of aid in many ways, without however, exerting control."[60] Compton complained to Bowman that there were "two attitudes of mind, the one conservative and the other adventurous," struggling for dominance in the profession. What Compton failed to note was that the "two attitudes of mind" often existed in the heads of the scientific leaders themselves.[61]

Compton's own "two-mindedness" was evident when he contemplated the idea of giving the government a role in funding new institutions at MIT.[62] Indeed, nearly all of Compton's efforts at MIT during the 1930s reflected the problems of trying to enhance the influence and prestige of scientific institutions without falling into dangerous dependency on either the corporate or the government sector. Upon taking over the presidency in 1930, Compton, with the support of Jewett and the other trustees, tried to alter the institute's image as a mere "technical" or vocational school for engineers. The new regime sought to provide greater emphasis on basic scientific research, in addition to providing first-rate engineering training. Most of the institute's income, however, came from companies that wanted to use the school to solve specific technical problems. Du Pont donated $1.1 million between 1911 and 1921, and George Eastman had given $10.5 million. In the absence of comparable alumni support or a spectacular endowment, these funds seemed to be a welcome source of revenue. But there were several problems associated with corporate support. Most of the problems posed by businesses involved technical issues of little scientific import. The volume of such work consumed a large percentage of the faculty's time, thereby making undergraduate instruction a distinctly second-class priority. When interesting problems were developed, companies often forbade the publication of any scholarly articles about them. But worst of all, industrial support was fickle and highly dependent on the vagaries of the business cycle. Academic support was usually high on a company's list of economy measures during hard times.[63]

The depression only intensified Compton's desire to weaken the institute's dependence on industry. But unfortunately for MIT, its other logical source of money, the foundations, were disinclined to give. They too were economizing during the crisis, and they simply assumed that industry would support the school. Compton was forced to look to the only other possible source of large-scale financial assistance: the federal government.

The school and the government had not been strangers to each other before 1930. Various small-scale technical projects had been carried out for the military and for federal agencies dating back to 1905, but Compton envisioned a substantial increase in the size and scope of the federal presence at MIT.[64]

In the early 1930s, he wanted the government to provide money to help develop a power transmission system that could be used by the Tennessee Valley Authority (TVA). Compton hoped for federal money to finance the creation of a large new laboratory at MIT where engineers would work on the creation of this proposed system. He also hoped research scientists in the new field of atomic physics might be able to work with electromagnetic accelerators similar to the one in the lab of Berkeley's up-and-coming scientist Ernest Lawrence. Both parties seemed likely to benefit from such a deal, but problems quickly arose that scuttled the endeavor. A third party had been introduced into the proposal, the Research Corporation of Boston, whose role was to manage any new patents that arose from the project, such as medical applications, which were anticipated to flow from the research in radiation that MIT hoped would be carried out in the new lab. Vannevar Bush, Compton's deputy at MIT, feared that the Research Corporation might charge such high prices for new medical applications that the school would be accused of being a party to medical price gouging. As difficult negotiations proceeded, Compton and faculty conservatives such as Dugald Jackson became increasingly suspicious of David Lilienthal's ideas about the social effects of the TVA. They were disinclined to get involved with what they perceived to be social engineering. By 1934, the parties were unable to settle disagreements about patent policies and the proposal collapsed.[65] The MIT–TVA–Research Corporation proposal would have involved MIT in a long-term project of financial and perhaps even political collaboration between the state, the company, and the school. Compton was unwilling to make that commitment in 1933 or 1934.

Thus Compton, like Jewett, found himself advancing contradictory positions in the 1930s: more science was needed to bring more prosperity, but more prosperity, as he acknowledged, often brought its own problems that required intelligent, scientific planning. Compton wanted to unleash the productive power of business, and he wanted the government's help in doing it, but he did not want the government to have any influence over the way the money it bestowed on scientists was spent. This contradiction was shared by many of the other younger leaders of the profession, and they could not formulate a politically acceptable way to receive government funding without meddlesome political interference.

The leaders of corporate science were ambiguously straddling two political worlds: a progressive liberalism that would have the state fostering scientific innovation, and a conservative emphasis on maintaining corporate power and protecting science from popular attack or political control. They were trying to achieve their goals while under attack from liberal politicians in the state and radical scientists from their own professional ranks.[66]

Karl Compton's arguments for greater funding played into that radical position: if science was indeed so central to the prosperity and progress of the nation, should not the national government play a responsible role in directing American science? But Compton believed that science could take care of itself and that conservative scientific leadership could take care of American science. Hence the political stalemate between the scientific profession and the Roosevelt administration in the 1930s. And hence also the inability of the profession's leaders to think their way out of their dilemma: how to get greater funding and greater political prominence for science and scientists while also keeping the state out.

That ideological stalemate would continue until world events made it possible for a new alliance between scientists and the state to emerge. The international crisis of the late 1930s provided an opportunity for Vannevar Bush, James Conant, and Ernest Lawrence to enact their own idea of a partnership between the profession and the state, and it would be one that avoided the difficult economic and political questions of the decade and emphasized instead the military usefulness of science. But before examining the wartime and Cold War careers of these three men, it will be useful to understand the personal and institutional factors that shaped their thought.

When war broke out in Europe in 1939, Bush, Conant, and Lawrence assumed the leadership of the nation's scientific effort, with Jewett and Karl Compton as well as Karl's brother Arthur serving as their subordinates. Bush was the most energetic and effective of those leaders, and his biography presents a vivid example of the way in which the dramatic institutional changes of twentieth-century life were creating opportunities for people from middle-class backgrounds to play significant roles in the nation's educational, corporate, and governing institutions. The style of Bush's scientific leadership also represented a sharp break from the conservatism of the preceding generation of corporate and scientific leaders.

Bush's father was a Protestant minister, and some of his New England ancestors had been whalers. He worked his way through Tufts University and after graduating in 1913 embarked on a series of professional affiliations that would take him through practically every outpost of American science: corporate laboratories, academia, independent invention, consulting, then a position in a major philanthropic organization and government advisory posts.[67] He worked briefly as a tester for General Electric at its Pittsfield and Schenectady facilities, a position from which he was dismissed along with an entire test crew after a fire. He next worked at the Brooklyn Navy Yard and then taught at Tufts. In the fall of 1915, he entered the joint MIT-Harvard engineering program and studied for a Ph.D. in electrical engineering.

He spent much of his career in academia as a professor and then a dean at MIT, but he by no means believed in the superiority of the academy over the world of commerce. He recounted in his memoir that his college instruction was generally of a poor quality; when he returned to graduate school he had to fight to get his degree in one year, which was as long as his money would hold out. His thesis, which he later admitted was "pretty crude," was supervised by A. E. Kennelly, "who was appalled when he learned that I proposed to finish in that year." Bush wanted the doctorate so he could begin a career as a consultant to industry. He also wanted to get married. Kennelly considered Bush "to be a heretic, and intended to stop what he thought was an end run." Bush thought his plan was working smoothly. He had "wisely based the theoretical treatment of [his] thesis on a branch of mathematics which Kennelly had never studied" and was nearly finished in the spring when Kennelly added to the required curriculum that he himself had approved for Bush's study.[68]

Bush appealed to Dugald Jackson, the chairman of the electrical engineering department at MIT. Jackson overruled Kennelly, and Bush was duly awarded his Ph.D. Bush developed a close relationship with Jackson, and the older man left an important, formative imprint on the young graduate. Jackson joined MIT in 1907 to take over the chairmanship of the school's electrical engineering department; he had himself worked as a consultant to the Sprague Electric Railway Company, the most important of the early electric streetcar companies, and later created his own consulting firm, Jackson and Moreland. He tried to elevate the engineering curriculum to the status of a first-rate academic discipline, and he advocated—over the strong resistance of his colleagues—the creation of a Ph.D. program in electrical engineering. Jackson was also responsible for reorienting the curriculum of MIT away from merely technical contracting work for industry and toward basic research. Jackson, in short, provided Bush with an immensely appealing ideal of the successful and practical professional man.

He was someone who, as Bush recalled in his autobiography, "was fully as proud of his contributions to the theories of electrical engineering as he was of being accepted as a sound consultant."[69]

But Jackson also embodied some of the worst pitfalls of the engineering profession's tightening connections to business. He believed that engineers had a moral obligation to defend utilities against regulatory attack, even if that meant producing scientifically dishonest data such as the type Jackson produced at the behest of the Philadelphia Electric Company in 1915, which the reformer-engineer Morris Cooke exposed.[70]

Bush, however, clung to the idealistic vision that Jackson espoused, and he rejected having his career completely subsumed in a corporate context. He argued against the idea that engineers should be nothing more than a group of "controlled employees," who had "no higher ideals than to serve as directed." This son of a clergyman saw his profession's task as that of "ministering to the people." But he was no bleeding-heart reformer. He was a staunch conservative who strongly concurred with Governor Calvin Coolidge's suppression of the Boston police strike of 1919. At the same time, he saw his profession as having a broader, more beneficent social role than simply that of increasing the wealth of the science-based corporations.[71]

This tension between professional integrity and public usefulness and accommodation to the existing order of things was a lifelong theme of Bush's career. He often seemed to be trying to steer a course between idealism and accommodation. Perhaps that tension explains the diverse character of his career before World War II. It was as if Bush was sampling all the different fields of American scientific practice—consulting, invention, academia—rather than settling on one in particular and facing the inevitable compromises that came with professional commitment.

He began his consulting career in 1917 and soon launched into a series of unsuccessful entrepreneurial ventures, one of which was the American Appliance Company, the original name of what is now the Raytheon Company.[72] He joined the engineering faculty of MIT in 1919, where along with Norbert Weiner he conducted pathbreaking work on an early version of the analog computer that Bush called a "differential analyzer." He rose rapidly through the ranks at MIT to become dean of the engineering school and ultimately vice-president of the institute. In 1936 he was appointed head of the National Research Council's Division of Engineering and Industrial Research. In 1938 he left MIT to become director of the Carnegie Institution in Washington, a post he would hold throughout his long career as a scientific adviser and administrative official in the government.

Upon becoming director of the Division of Engineering and Industrial

Research in 1936, Bush sought to alter the composition of the division so that it might include more engineers with experience in large corporate enterprises—men who, Bush believed, would be of service to the nation in the event of an "emergency" such as the one of 1917–18.[73] "Historically," he wrote to his colleagues in the NRC, "the Division has been largely made up of representatives of various societies selected by those scientists themselves. This through no fault of the societies or the individuals involved, has resulted in a membership which cannot be regarded as the best possible selection of individuals for the purposes in mind." To attract the best men, a completely new set of standards needed to be applied. He wanted "some men who know the history of the Division and can carry forward its traditions." The balance of the list should preferably be younger men "in key positions in the industrial and engineering life of the country, capable of acting effectively and energetically in an emergency on the type of problem which will then be before the Division." The concerns of such a group should be equally divided between the research needs of government and industry—and the relation between the two.[74]

Bush explained to hostile and resistant academy members who considered engineers second-class scientists that "some engineers develop in the direction of research and others in the direction of operating executives. Men in the latter group may be even more valuable in an emergency than the research or academic men." A compromise was reached in September 1938. A revised set of bylaws created three categories of members, consisting of people from the engineering section of the NAS, members of the engineering societies in the Division of Engineering and Industrial Research, and a category called "Members at Large," who were to be chosen by the president of the division in consultation with the presidents of the societies. This change allowed Bush to appoint the corporate engineers he wanted.[75]

Bush's battle against the stodgy membership policies of the NAS not only shows how early he was concerned with linking the scientific and engineering professions to the military and the state but also illustrates a larger motif in his career. In his rise to success and to positions of influence in various institutions, he struggled continually against the snobbery of the "traditional" elites who dominated the fields in which he worked. Years later, his memoir bristled with tales of condescending slights inflicted by people like Cameron Forbes, a trustee of the Carnegie Institution and a descendant of the "Boston men of affairs who molded the nineteenth and much of the twentieth century." Forbes's father had been one of the original organizers of AT&T "and Cam never let anyone forget it." Bush described how Forbes "kept a fine stable of horses, played match polo until he was sixty, furnished mounts to the Harvard polo team . . . and drove a coach and four through narrow crooked roads" on his Georgia estate. He did this, Bush was convinced, to "show me he could, and to impress upon me that I could not."[76]

Shortly after taking over at the Carnegie Institution, Bush wrote, "a minor battle [with Forbes] was needed before it became clear who was running the show." Forbes tried to assert his sense of superiority in juvenile ways: for example, Bush found himself humiliated when Forbes decided to monopolize the time of one meeting of the institute's Executive Committee by criticizing the grammar in one of Bush's reports. "Fortunately," Bush recalls, "I was sitting beside Walter Gifford, then president of AT&T, and as my steam pressure was reaching the explosion point, he whispered in my ear, 'Keep your shirt on, Van; he does the same thing to me.'" His recollection of this incident nicely illustrates his simultaneous resentment of elitist snobbery and his pleasure at being accepted by other members of the elite.[77]

Bush took pride in his ability to gain entry into elite circles through his intelligence, energy, and accomplishment. But he recognized that there was a question of political and even cultural accommodation involved. He noted that Cameron Forbes belonged to a "tribe [that] felt it had the right, and also the duty, to run everything in sight." Bush was disturbed by "those who felt strongly that they should carry on in the spirit of their ancestors, even though some of the ancestors had not enhanced the public's welfare appreciably. They accepted into their circles outsiders whose ancestors had not done anything striking, but only after testing them to be sure they should not be thrust into outer darkness."[78]

Bush noted that his association with Forbes at least had the redeeming feature of allowing him to meet Herbert Hoover. They met at Forbes's house one Sunday morning while Hoover was working on a speech; the last part of the speech was giving Hoover trouble. He asked for Bush's advice and even adopted one of his suggestions. Bush recalled fondly, "No man with an inflated ego ever does that sort of thing with a younger man, and very few men who have held the greatest posts in the country would come anywhere near it. I promptly joined the ranks of the considerable company who were proud to call him 'Chief.'"[79]

Hoover was someone who dwelt in the same ideological territory in which Bush found himself. He was a successful businessman but also a public-spirited one. Hoover's ideas about the role of the engineer in America, which he had been trumpeting throughout the 1920s, were an attempt at resolving the ideological contradictions embodied and left unresolved in the thought of engineers and scientists like Jewett: how could American corporate scientists maintain their professionalism and idealism while working within the corporate system? The engineer was clearly tied to the business system, Hoover argued, and this was not something to decry. But Hoover also believed that the engineer had an obligation to a larger, broader conception of the public good, which had to be upheld even when corporate interests threatened it.

In the 1920s, Hoover spoke of the necessity of steering an economic course in America that was "based neither on the capitalism of Adam Smith nor upon the socialism of Karl Marx." Engineers had a responsibility to help steer the business system back toward that middle course when it went astray. They had "to visualize the nation as a single organism and to examine its efficiency towards its only real objective—the maximum production." Hoover also stressed the importance of achieving a classless meritocracy in America, an idea that would be very important in the later careers of Bush and James Conant. There must be no inherited aristocracy in America, he argued in his 1922 book *American Individualism*. There must be "equality of opportunity" and avenues through which people could "rise from the mass."[80]

Hoover's "middle-way" approach to the role of scientists in America's political economy was an important idea for Bush. The effect of Hoover's ideas, as well as those of Walter Lippmann, come through vividly in a speech Bush delivered in January 1939, "The Qualities of a Profession." In it he acknowledged that the relatively new engineering profession derived its existence from the confluence of business and science. But he hoped the profession would develop an ethos of disinterested public service. No profession, he noted, had been able to maintain its legitimacy in the eyes of the public without exuding a clear sense that its fundamental purpose was that of "ministering to the people."[81]

There was a strong basis for hoping that the status of engineers would rise, Bush noted, since the business culture itself was developing a more truly professional ethos. "One of the most encouraging signs of the times," he wrote, in a near perfect echo of Lippmann, "is the gradual emergence in our day of the truly professional man of business. This is occurring for one reason[:] because of a gradual change in corporate form." Management had become a "self-perpetuating entity" whose responsibility it was to "minister" to the owners, the workers, the government, and the consumers. Instead of mere profit-seeking and competition, business professionals had made a commitment to "managing prosperity so that it will be conducive to the health of a nation."[82]

This view of professional culture, Bush made clear, was one that transcended and obliterated political conflict. Engineers, working within the business system, replaced "the hopelessness of Malthus with an embarrassing plenty." They needed to make sure that they oriented their work "primarily toward the advancement of public health, safety, comfort and progress." If not, they would have failed as public figures, and "we may as well conclude that we are merely one more group of the population . . . maintaining our economic status by a continuing struggle against the interests of other groups . . . with no higher ideals than to serve as directed, and with no greater satisfaction than the securing of an adequate income as one member in the struggle for the profits of an individual age." The profession's legitimacy, in short, depended directly on the extent to which engineers could transcend the self-interested conflict that characterized the American commercial marketplace and collaborate with other elites. In another revealing piece written at the same time, Bush noted that "the professional man must learn to operate more effectively as part of a complex organization while still retaining his professional status," and the professional's role must always be one of "detachment" and acting without regard to the strife around him. Such a conception of his professional and political role would strongly influence Bush's conduct as a scientific administrator and public figure in the 1940s and 1950s.[83]

In 1939 and 1940, Bush received crucial help in his efforts to mobilize America for war from James Conant, a chemist and the president of Harvard. Conant, who was also a product of a middle-class New England family and who, like Bush, was a moderately conservative Republican, rose to the presidency in 1933 at the age of forty. The early years of his tenure were notable for instituting reforms that broadened the student base, while also making it easier to dismiss non-tenure-track faculty. When the contracts of two leftist professors were not renewed, the dismissed teachers accused Conant of purging them for their political views.

Perhaps because of the acrimony that resulted from that episode, the young president was otherwise circumspect about becoming involved in political controversies. This caution could occasionally go to absurd and even shameful lengths, such as the time Hitler's press secretary and Harvard alumnus Ernest Hanfstaengl ('09) came to the twenty-fifth reunion of his class and brought along DM 2,500 for the purpose of creating a scholarship in his name. Conant rejected the scholarship offer but did not speak out as forcefully as some hoped he would against the spectacle of a Harvard alumnus giving Nazi salutes at the class day parade and even inspiring the younger alumni of the class of 1924 to break into a goose step and give their own Heil Hitler salutes to the crowd.[84] With regard to the loyalty oath imposed by the Massachusetts state legislature in 1935, Conant put conformity before principle. While expressing disapproval of the idea of a loyalty oath, he publicly announced his willingness to submit to it once it was enacted, and he demanded that all Harvard faculty follow suit under penalty of dismissal.

Notably, the only time Conant took a decisive and controversial stand on a matter of intense public controversy was when he aligned himself with the advocates of American intervention in World War II during 1940 and 1941.[85] He was a member of William Allen White's Committee to Defend America by Aiding the Allies, and at the 1940 commencement he was booed by the graduating class, many of whom were angered at the example of a comfortable, middle-aged academic man urging younger men to risk their lives in a war that many regarded as a matter of no concern to America. But such criticisms did not deter him. Conant noted in his diary in June 1940 that he had "expressed my views [to the president of Princeton] on U.S.A. armed to the teeth, belligerent and running the world." He envisioned a "Pax Americana like the Pax Britannica of the 19th century."[86]

Conant's attitude about waging war and about the involvement of scientists in military matters had been shaped by important and tragic personal experiences during World War I. Organic chemists like Conant had come into demand once war broke out in 1914 because the supply of chemicals from German manufacturers had been cut off. Conant and two other Harvard graduates decided in 1917 to form a business manufacturing pharmaceuticals, but the enterprise was undone by failure and disaster. The first space the three young chemists rented was destroyed when their operation set the building on fire. A greater disaster occurred when a second plant was obtained. Their manufacturing procedure had not been sufficiently thought through; a chemical reaction took place too quickly and an explosion occurred, burning the plant to the ground and killing one of Conant's partners, as well as two workmen.[87]

Conant was not present at the time of the disaster. The company was liquidated and the stockholders paid off from the royalties of a patent based on Conant's work. "This tragic experience with applied chemistry should have discouraged me for a lifetime," Conant wrote. "In fact it did not; within eighteen months I was to become involved again (as a chemist in military uniform) in developing a new manufacturing process for a certain chemical—this time a poison gas," lewisite.[88]

Having failed as an entrepreneur, Conant found his career boosted by more direct service to the needs of the military: he became an officer in the Chemical Warfare Service. He had been recruited in 1917 to lead a research group at the American University in Washington, D.C., under the auspices of the Bureau of Mines. In September 1917 he received a commission as a lieutenant in the Sanitary Corps. In 1918 he was promoted to a major of the Chemical Warfare Service and transferred to Ohio, where he worked on the production of lewisite. The experience was, as Conant's biographer writes, "a low-tech preview of Los Alamos."[89] His defense of the experience also shows his lack of compunction over the use of his expertise in the manufacture of weapons. "To me," Conant wrote during the Vietnam War, "the development of new and more effective gases seemed no more immoral than the manufacture of explosives and guns. Tear gas for the use of police forces was a gift to the forces of law and order from the Chemical Warfare Service, so to speak. I did not see in 1917, and do not see in 1968, why tearing a man's guts out by a high-explosive shell is to be preferred to maiming him by attacking his lungs or skin."[90]

The other chemists with whom Conant served in Willoughby, Ohio, were eager to obtain positions in industry. They tried to persuade Conant to abandon his academic career and join the ranks of the corporate scientists. They argued that industrial work brought more tangible satisfactions; one could see the fruits of one's work rolling off the production line. Conant resisted their urgings. But he also moved beyond the standard academic prejudice against the applied scientist. His disastrous experience as an entrepreneur, as well as his work in the Chemical Warfare Service, had altered his point of view: "The ideal scientific investigator no longer stood alone on a high pedestal. I recognized that there were other fascinating ways of using one's energies. I cherished a carefully hidden ambition to try my wings someday in other fields than the scientific." Conant did maintain some professional connections with the corporate world even if he did not fully join it. For example, he supplemented his Harvard income and the financial resources that derived from his inheritance with consulting fees from Du Pont and from the American Petroleum Institute. Conant easily accommodated himself to the new professional-political culture of the 1940s, in which scientists from many different disciplines and institutions worked together on military technologies under the organizational umbrella of the state.[91]

The other academic scientist to emerge as an important institutional leader in the late 1930s was Ernest Lawrence. Lawrence's great achievement, his cyclotron, which he expanded dramatically in size and electrical power throughout the decade, offered an innovative mixture of science and technology and was one of the original examples of group research, or big science, conducted in an American university. The steady expansion of the cyclotron and of the Radiation Laboratory that Berkeley provided Lawrence required that he establish connections with the dispensers of wealth in the nation's largest philanthropies.[92]

To carry out his continuous technical expansion, Lawrence undertook a campaign of fund-raising throughout the 1930s. These forays brought him into increasing contact with wealthy investors and philanthropic institutions, many of whom were won over by Lawrence's infectious high energy, his very worldly pragmatism, and his eagerness for success. His efforts were capped by an award of $1.15 million from the Rockefeller Foundation to build an accelerator that would reach the 100-million-volt level.[93]

Lawrence's lab and his personality in those early years in the 1930s reflected an irresistible exuberance. The press responded to this quality of his work and lavished attention on him as a young, enterprising American hero, forging new paths of knowledge and finding a practical application for that knowledge: the production of radiation isotopes for cancer treatment. Lawrence also employed the tactics of the vaudeville showman to demonstrate the medical potential of his cyclotron's radioactive products. Radio-sodium was one such product. It provided the same therapeutic effects as radium but had none of the latter's harmful consequences. He regularly demonstrated its use before public audiences with a stunt: a volunteer from the audience (who, in one case, was Robert Oppenheimer) was brought onstage and asked to drink a radio-sodium mixture; a Geiger counter was then passed over the straight man's stomach and then to other areas of the body to demonstrate the diffusion of the isotope through the bloodstream.[94]

Radioactive isotopes for cancer treatment gained the lab widespread and positive press attention. John Lawrence, Ernest's brother and a practicing physician, used the lab's products in treating his own patients. In a melodramatic episode, the Lawrence brothers healed their own mother successfully with a Sloan X-ray tube. After having been diagnosed at the Mayo Clinic with terminal cancer of the pelvis, Gunda Lawrence survived the radiation treatments administered by her sons; her cancer disappeared and she lived eighteen more years.[95]

The medical use of radioactive isotopes and neutrons became increasingly important to the lab during the 1930s as possible sources of funding began to dry up. In a letter to Niels Bohr, Lawrence felt the need to "confess that one reason we have undertaken this biological work is that we thereby have been able to get financial support for all of the work in the laboratory. As you know, it is much easier to get funds for medical research."[96]

The lab's growing success and fame cast some slight shadows over an otherwise brilliant scene. The growing size of the magnet and the cyclotron took on a crude, symbolic power: everyone in the lab was inexorably, sometimes exhaustingly, drawn toward its operation. All were devoted to Lawrence's single-minded and somewhat stereotypically American determination to make the thing bigger and more powerful, to increase its capacity for generating wealth and fame, without thinking very deeply about the professional or social consequences of this kind of scientific work.

Robert Wilson was one who tired of the obsessiveness and fled to Princeton, where he was free once more to pursue his own research. "The sense of history being made" at the lab did not outweigh the sacrifices to his own scientific ambitions. The relentless concern with increasing the power of the accelerator was, Wilson judged, "an activity that epitomized team research at its worst." Even sharper criticism would come later from Lawrence's boyhood friend Merle Tuve, who worked in the Carnegie Institution in Washington in the 1930s and 1940s. Tuve saw "big science" as a threat not only to the professional culture of physics but, in a subtle way, to the spirit of American democracy itself. By contrast, Tuve's wartime work on the proximity fuse was a refreshing, small-scale effort that exemplified the "rediscovery . . . of the efficiency of the democratic principle of directing the effort of an organized group of people." In contrast to the big-science approach of "just giving orders from above," the more democratic, small-scale approach was one in which "criticism flows both ways," and all involved are allowed to feel they are making a significant contribution to the enterprise.[97]

Lawrence was mindful of this problem, and in 1937 he commented that he himself "did not even know what substances are being bombarded or exactly what is being done." He acknowledged that "there is always an urge to do what seems to be the obvious thing to make the cyclotron work better, at the expense of actual nuclear research." While the cyclotron could exert crude power over the Radiation Laboratory, it was being depicted in the media in even cruder terms. The most common analogy was that of a weapon—the atomic machine gun. Donald Cooksey, Lawrence's principal assistant, told a reporter at the 1939 Golden Gate International Exposition that the cyclotron could shoot more "atomic bullets" than "one gun could fire in three million years." The press referred to it as artillery, or the "big guns." In a few years' time Lawrence's lab would be deeply immersed in work on atomic ordnance with a destructive power that not even the most sensationalistic journalists envisioned back in 1939.[98]

The institutional power of this new generation of leaders was apparent when the nation began to mobilize its scientists for war in 1939–40. Jewett sought to create a prominent role for the National Academy of Sciences in the mobilization effort, but his proposal lost out to a more ad hoc organization proposed by Vannevar Bush. Jewett's proposal was presented in a way that clearly reflected his hostility toward the Roosevelt administration. In a vehement 1941 essay coauthored with an AT&T associate, Robert W. King, he argued that the government's lack of preparedness spoke to a more pervasive failure on the part of the Roosevelt administration properly to use expert knowledge for the solution of social problems. What was needed, Jewett wrote, was some concrete, institutionalized method for bringing scientific and technical expertise to bear on national problems. A "new instrument of inquiry and investigation" was required so that government "can educate itself 'to act with vigor and economy.' "[99]

Corporate science as embodied in the National Academy of Sciences, Jewett believed, suggested a way. The pooling of scientific expertise and the group approach to problems could be applied to politics. "Science," King and Jewett wrote, "has forged ahead in the domain of organization and has succeeded in creating a sort of superbeing which is vastly more effective than the unit individual." The public had overcome the hostility to science that was characteristic of the early 1930s, and it now appreciated the "pragmatic success to which science and organized knowledge have attained." It recognized that scientists (unlike politicians) stood for the proposition "that what counts in the world today is accuracy and truth, not guessing." The public, in short, was "ready to accept a wider application of the scientific method—or the nearest approach to this method which is practicable in the affairs of state."[100]

The current ramshackle quality of democratic governance, as practiced by the Roosevelt administration, had to be replaced, he argued, by a more frankly elitist structure. Jewett noted that the Framers had never intended the country to be a democracy but had designed the government so that the best and most effective citizens could act on behalf of the public good. He buttressed his argument in favor of what Ronald Tobey referred to as a "scientific oligarchy" by quoting James Madison: "Democracies have ever been spectacles of turbulence and contention and have ever been found incompatible with personal security and the right of property, and have in general been as short in their lives as they have been violent in their deaths."[101]

Jewett's sharply elitist conception of his role was hardly one that the Roosevelt administration could embrace. He acknowledged that Bush would be a more effective scientific adviser to Roosevelt even though in a strict statutory sense Jewett, as president of the NAS, held that role. By 1941, Bush was creating just such a post for himself. Through such organizations as the National Defense Research Council and later the Office of Scientific Research and Development, Bush would spend the war years acting as the American scientific profession's ambassador to the Roosevelt White House. Those two organizations were the first of many that would transform the culture of the scientific profession and the state itself.

American science and, indeed, American political culture as a whole survived the "legitimation crisis" of the 1930s. Both American science and American capitalism had come under attack, but their prestige remained intact. Philanthropic dollars continued to flow into the coffers of researchers like Ernest Lawrence; science-based companies escaped any serious or injurious regulatory attack; and if attendance at science museums and world fairs were even a crude gauge of public attitudes toward science and technology, it seemed the public retained a mixture of fascination about current marvels and an expectant optimism about what the "world of tomorrow" might bring.

The fundamental ideological character of both the profession and the state had not changed very much as a result of the economic crisis: American scientific leaders were still profoundly influenced by an evolutionary paradigm according to which ever more progress would occur as long as money continued to flow for scientific research. The nation's political elite, in contrast, was still dedicated as it had been before the depression to the protection and cultivation of what Louis Galambos has described as the "corporate commonwealth."

But both the scientific and the political elites had gone through an important process of ideological "filtering." Scientists who offered themselves to the public as stern exemplars of fundamental conservative "truths" about social as well as natural phenomena were either directly refuted or were quietly superseded by institutional interests and organizational imperatives that sought to mute any political harshness to the role of science and technology in American life.

The outbreak of World War II found the American scientific profession under the leadership of men whose experience and professional philosophy had been shaped in the new organizational and cultural context of corporate science. Jewett led the National Academy of Sciences (his election was the result, in part, of strenuous maneuvering by Bush); Bush headed the Carnegie Institution of Washington; Lawrence was the director of the nation's, if not the world's, premier nuclear research center; Conant and Karl Compton were the presidents, respectively, of Harvard and MIT. They were people who strongly believed that scientists and engineers could work together in large organizations and prove themselves valuable partners to the leaders of American business, as well as to the nation's political elite.

Among that group, as we will see in the next chapter, the most influential in shaping the new alliance with the state were Bush and Conant. Both of these men had spent their professional lives primarily in the academic sector, yet they adopted many of the political attributes that characterized the culture of corporate science. Like Jewett, Whitney, and Kettering, they believed basic research and scientific innovation were central to American prosperity. They likewise believed that elite scientists had important roles to play as cooperative, politically neutral partners with the powerful men who ran the elite institutions of America, and they were willing to cooperate with the government on its own terms and leave the politics of the Roosevelt administration in the hands of the president and his men.[102]

They would perpetuate many of the qualities of corporate science in their wartime work, and they would implement many of the political and professional ideals of that culture. But wartime science would provide its own set of distinctive political realities and problems. The persistence of older professional patterns in new institutional and political settings would provide many interesting tensions in the careers of Bush, Conant, and other leading state scientists in the 1940s.

Some bitter schisms would ultimately result within the ranks of the scientific elite, but the political and institutional experiences of the war years would be assimilated into an ideological framework that had been built by the corporate scientists of the 1920s and 1930s. The clash between older values and new experiences would also lead to new ideologies and important changes in both American science and American politics. Karl Compton provided a glimpse of the professional ideology that would emerge during the 1940s when he wrote in 1941 that basic research in the war effort was "yielding new developments, new techniques and new understandings which will have important peacetime applications and which presage a new prosperity for science and engineering after the war."[103]

The idea that science and scientific expertise were essential to national progress and prosperity was greatly intensified when scientists began to argue, and politicians and military officials began to believe, that science was also critical to national defense. When notions of evolutionary development and expert scientific guidance began to be applied to the nation's military arsenal as well as to the national economy, the professional philosophy of the state scientists would become freighted with ominous political significance.

— — — — — — — — — — — — — — — — — — — — — — — — — — — — — — — — — — — — — — — —

## CHAPTER 3

## THE CREATION OF STATE SCIENCE

— — — — — — — — — — — — — — — — — — — — — — — — — — — — — — — — — — — — — — — —

Scientists got out of the engine room and onto the deck of the ship of state in the 1940s. Once they got there they created a new professional and public role for themselves, with nearly all of the leading scientific administrators of the 1930s serving as advisers to and collaborators with the military and the political elite. They created this new realm of state science by using many of the ideas and organizational patterns developed from within the cultures of corporate science during the 1920s and 1930s.

The conflicts that emerged in state science during the war years revealed the extent to which many of the ideological and professional conflicts that characterized corporate science were continued into the 1940s. Scientists continued to disagree among themselves about their public role, their relationship with the government, and the fundamental question of who should control and manage the creations of science: scientists or politicians. These differences were revealed in many episodes during the 1940s such as the arguments over whether to proceed with a fission bomb project. Once that issue was answered affirmatively, the terms of collaboration among scientists, military officials, and political leaders were fought out in the administrative and technical problems of the project and also in the momentous political question of whether to use the atomic bomb against Japan. After the war the debates continued in the discussions over the perpetuation of the wartime institutions.

The professional and ideological vision that prevailed was the one most closely associated with the ideas of Vannevar Bush and James Conant, and their ideas prevailed for many of the same reasons that such leading corporate scientists as Carty, Jewett, Whitney, and Kettering were able to win influential positions in American industry: their ideas promised to enhance rather than usurp the powers of their benefactors. They promised a partnership that would benefit science, the state, the military, and the public through an alliance of collaborating elites who administered the key institutions of the country in a way that was meritocratic and flexible. According to their vision, these institutions would respond to national needs in times of crisis and foster a political economy of innovation and abundance in times of peace. By the end of the 1940s, the economic promises of corporate science would be tied to a militaristic sensibility that emerged during the war years to create not only a new public and professional role for scientific expertise but also a distinctive new governing ideology: scientific militarism.

The road to that new ideology was neither inevitable nor unobstructed. But from the conflicts over the terms of collaboration between science and the state emerged important changes in American political culture: a greater acceptance of the role of expertise in shaping policy; a greater commitment to meritocracy; and a tightening of interests among the military, professional, political, and corporate elites.

Some of these changes had been brewing since 1917–18, when science was first mobilized by the state during wartime. Reformers seized the war that America entered in 1941 as an opportunity for transforming the state and the political culture, just as the opportunity of 1917–18 had been. In both instances reformers wanted to expand the administrative capacities of the state and to carry out those treasured Progressive goals of greater cooperation and efficiency. But there were important ideological differences as well between the experiences of these two generations.[1]

The contrast is similar to the change from the elitist reform agenda of Henry Towne to the goals of corporate scientists like Carty and Jewett. The 1917 generation saw their war as an opportunity for moral reform and for the strengthening of a static class system. Above all, they saw their measures as a temporary experiment. But Bush, like the corporate scientists, eschewed moral reform. He was more interested in perfecting the governing system of the nation than in improving the mores of soldiers or citizens. And he wanted his changes to be permanent.

Neither Bush nor any of the wartime administrators of American science in the 1940s could have said anything comparable to what Leonard Wood said in 1916 during the preparedness controversy. Military service, Wood argued, would create "a more valuable industrial factor because of his better physique, his improved mental and physical discipline, and greater respect for the flag, law and order, and his superiors." It would reduce crime and bring about "the moral organization of the people: an organization which creates in the heart of every citizen a sense of his obligation for service to the nation in time of war or other difficulty."[2]

For men like Bush and Conant, the locus of political reform had shifted away from the citizenry and their physiques and toward the pattern of institutional relationships that structured and governed the experience of the citizens. This emphasis on institutions was accompanied by the creation of a new professional sensibility among the leading scientists, one that stressed their practicality and their ability to serve as effective—but not threatening—partners with the military and political elite.

Both during and after World War II, Bush, Conant, Arthur Compton, and J. Robert Oppenheimer championed this vision of scientists as collaborating partners with the powerful. Like the corporate scientists of the 1920s and 1930s, they resisted the idea that their work had any political implication. They did not aspire to be a special interest that would be at odds with the military and political leaders of the state. They explicitly rejected the efforts of other scientists such as Leo Szilard and James Franck, who tried to assert a distinctive, adversarial policymaking role for scientists. Bush and Conant and the other leading wartime administrators were the "winners" of the political conflict that unfolded during the 1940s, and while they will be at the center of the story told in this and the following chapter, it is important to keep in mind the presences of dissidents like Szilard and Franck, for they constantly challenged the notion of compliant collaboration with the corporate as well as political and military elites. The alternative vision offered by the dissident scientists did not prevail, but it did not completely disappear either.

The very first place where we can see the new, collaborative sensibility emerging—and being challenged—is in the origins of the Manhattan Project. Jewett, Bush, and Conant were reluctant participants in the initiation of the project. Jewett, who was a division director in the National Defense Research Committee (NDRC), supported Bush's skepticism about the wisdom and the professional implications of pursuing the then unlikely possibility of an atomic bomb. The two men's correspondence in 1940 and 1941 reveals the extent to which Bush wanted to impress the military with the reliability and sound judgment of the scientific elite. Bush's mission also closely conformed to those of a corporate scientific administrator such as Jewett had been at AT&T: that is, he had to manage a scientific enterprise that included scientists eager to engage in basic research, while also ensuring that applied scientific "products" (weapons) were made available to his employers and superiors. Like Jewett at AT&T during the 1920s, Bush tried to show that scientific administrators were as crucial to a warring nation as they were to a monopolistic business.

The most salient and perhaps most significant aspect of the professional role Bush was proffering was its apolitical character. This was evident when this lifelong Republican got Harry Hopkins to help convince Franklin Roosevelt to create the NDRC in June 1940. Bush repeatedly stressed his ability to work with Hopkins and FDR in total disregard of their profound political differences.

Once the NDRC started operating, he also wanted to overcome the stereotypes that military people harbored toward his profession. In the interwar years, Bush noted, "military laboratories were dominated by officers who made it utterly clear that scientists or engineers employed in these laboratories were of a lower caste of society." Scientists were forced to work under the strict control of military officials whose scientific knowledge was "rudimentary at best." To the military, "an engineer was primarily a salesman, and he was treated accordingly."[3]

To Bush, the ideas being bandied about concerning a possible fission bomb were just the kind of disastrous venture that would alienate the military. But a growing contingent of American and European physicists, among them some of the world's most famous, including Ernest Lawrence, Arthur Compton, Leo Szilard, and even Niels Bohr, were talking energetically among themselves about the feasibility of such a weapon. They waited anxiously for the American government to move on the atomic project, but nothing happened throughout 1940 and well into 1941.

Bush's NDRC colleagues James Conant and Frank Jewett shared his emphasis on practicality. They too tended to act and think with the cost-conscious discipline of a corporate research project leader. The fission proposal struck all of them initially as an academic research fantasy that could not be indulged amid the intense pressures of wartime production. Roosevelt did not seem to feel any urgency either. He referred the matter to Lyman Briggs, the chief physicist at the Bureau of Standards, who formed the Uranium Committee of advisers.[4]

In May 1940, Bush told Frank Jewett that he had "hoped last summer that we had heard the last of atomic explosions from uranium fission, and I wish we had." But Bush was "puzzled as to what, if anything, ought to be done in this country in connection with it." It seemed like a terrible long shot, but Bush also realized "that one is not likely to know what others are doing."[5]

Bush's desire to stress the practicality of the American scientist was tempered by his realization that he did not fully understand the science behind fission. He admitted in his autobiography that he was out of his depth with regard to the technical issues involved. "At times," he wrote, "I have been called an 'atomic scientist.' It would be fully as accurate to call me a child psychologist." He admitted that he "made no technical contribution whatever to the war effort." His function was first and last administrative and political. "I regarded myself as a link between the President and American science and technology," he wrote, "and not as an oracle or an expert on all matters scientific."[6]

Bush may not have understood the science behind the hypothetical fission bomb, but his language and his pattern of thought continuously fell back on the example of a corporate research manager. He felt the fission problem amounted to a business manager's decision about the commitment the uranium effort should receive. Bush wrote Jewett, who as a cost-conscious corporate scientist was likely to be sympathetic, that "it could be possible to spend a very large amount of money indeed, and yet there is certainly no clear-cut path to defense results of great importance lying open before us at the present time."[7]

Bush's NDRC in 1941 continued to adhere to a strict emphasis on technologies that had immediate applicability. Conant strongly concurred in this approach. He had traveled to England on NDRC business and had experienced a few German air attacks. He returned shaken by the gravity of the situation and was determined that nothing should prevent the United States from providing help to Britain. "To me," he later recalled, "the defense of the free world was in such a dangerous state that only efforts which were likely to yield results within a matter of months, or, at most, a year or two were worthy of serious consideration." He "was impatient with the arguments of some of the physicists," and their enthusiastic talk about the civilian as well as military potentials of fission "left me cold. I suggested that until Nazi Germany was defeated all our energies should be concentrated on one immediate objective."[8]

In the spring of 1941, nearly a year after expressing his concerns to Jewett, Bush had still taken no actions toward a fission bomb. Ernest Lawrence began to act unilaterally to push Lyman Briggs and Bush to greater efforts. He made use of his personal connection with the investment banker Alfred Loomis, who had been a patron of Lawrence's and who was also a cousin of Secretary of War Henry Stimson. Lawrence lunched with Loomis and Karl Compton and made his case. An inspired Compton then called and wrote his former colleague Vannevar Bush to convey Lawrence's complaints.

Compton argued that the project needed someone like Lawrence because Briggs was "by nature slow, conservative, methodical and accustomed to operate at peace-time government bureau tempo." The "ablest theoretical physicists" in the country had to be brought into the project, as had been done in the radar program at MIT. Compton finished by suggesting that Lawrence be made Bush's deputy. "I believe that the results," Compton concluded, "would be decidedly interesting."[9]

They were, but not in the way Compton anticipated. Bush was furious.

He resented the challenge to Briggs's competence and, by not so subtle implication, to his own. "I think Briggs has done exceedingly well to keep his balance, and to approach the matter on a basis which would seem to me to have great sense," Bush wrote to Jewett. "Moreover," he added, "I think that Briggs is a grand person to have in the matter, and I have backed him up to the best of my ability and I intend to do so in the future."[10]

Lawrence, Bush wrote dismissively, was "an enthusiastic physicist who has not worked in harness too much." He was not, in short, sufficiently "practical" and had not been harnessed by the disciplines that tempered engineers and corporate scientists. He was particularly angered by Lawrence's use of his connections to apply pressure. As Bush recounted it to Jewett, he told Lawrence in a meeting "that I was running the show, that we had established a procedure for handling it, that he could either conform to that as a member of the NDRC and put in his kicks through the internal mechanism, or he could be utterly outside and act as an individual in any way he saw fit." Bush replied to Compton that he would suggest to Briggs that Lawrence be made Briggs's assistant. Lawrence, Bush wrote, "got into line. . . . I think this matter was thoroughly straightened out . . . but it left its trail behind."[11]

Bush was doing more than simply protecting his turf and his ego. He was trying to lay down the law as to the style in which state science was going to be practiced: hierarchical, pragmatic, with clear goals and clear lines of authority, with research projects undertaken only with the consent of the highest organizational authorities—rather like the corporate research laboratories in the 1920s and 1930s.

In mid-1941 Bush slowly began to respond to the arguments made by Lawrence and also by British scientists. But the emphasis should definitely be placed on the word "slowly." The nature of that resistance is revealing. Bush would not act unless the fission project won the wholehearted support of the nation's corporate scientists. Under growing pressure to increase their activity, both Briggs and Bush realized that their ignorance of the subject prevented them from making an informed determination about its feasibility. They asked Jewett to form an NAS committee to investigate fully the plausibility of some military application of nuclear fission. The committee that Jewett appointed consisted of Arthur Compton as chair and William Coolidge as cochair. Coolidge had recently retired as the director of GE's research lab. The other three members were Lawrence, J. C. Slater of MIT, and H. J. Van Vleck of Harvard. Jewett had also wanted his old colleague Bancroft Gherardi to serve on the committee, but Gherardi was unable to do so because of illness.[12]

This was to be the first of three NAS committees to study the feasibility problem. This first body issued a report on May 17, 1941. It proposed three possible applications of fissionable uranium: radioactive materials "scattered over enemy territory," a power source for submarines, or "violently explosive bombs." Jewett transmitted the report to Bush with a letter in which he described it as "authoritative and impressive," but he also expressed acute sensitivity to time and cost. He said he had a "lurking fear" that the report "might be over-enthusiastic in parts and not so well-balanced."[13]

Jewett had spent a professional life making judgments similar to the one Bush was now forced to make. The corporate scientists had combined the intellectual curiosity of the basic researcher with the cost consciousness of the businessman; they took chances on new innovations that might prove useful and lucrative, but there had to be a minimal correspondence between cost and potential usefulness. Jewett's business instincts told him that the costs were far greater than the potential results. "It would be foolish," he wrote to Robert Millikan, whose advice was also sought on the feasibility of the weapon, "to proceed solely on the basis of one-sided enthusiasm and a trust that in an eight-handed poker game the Lord will always enable us to draw the right two cards to complete a royal flush." Millikan seconded Jewett's caution. He noted that "conservative administrative procedure would seem to require concentrating upon the settlement of" the chain reaction problem before going ahead with any full-scale effort. But Millikan doubted that anything militarily useful could ever come of such research. "The problem should not have an A-1 priority for the purposes of this war."[14]

Bush was not entirely heartened by Millikan's views when Jewett gave him a copy of the letter. He felt that "a very long period of engineering work of the most difficult nature" would be needed before "anything practical can come of the matter." There needed to be greater input from engineers and less hot air from the likes of Ernest Lawrence. Toward that end, he and Conant asked for a second NAS report, and this time Bush insisted on a greater presence of corporate scientists on the panel. Coolidge of GE was asked to chair the committee instead of Arthur Compton. The new members were Oliver Buckley of Bell Labs (Jewett's deputy) and L. W. Chubb, the chief research scientist for Westinghouse. Lawrence was not on this committee.[15]

The second NAS report, submitted on July 11, 1941, was optimistic about obtaining "useful power" from uranium fission, and it recommended the establishment of a central laboratory for work on the chain reaction, not surprising for men used to centralized scientific work. The report also advocated the allocation of more money: $250,000 per year and an additional $500,000 to $1 million for materials. They concluded that the possible military applications justified an intensive effort. But as Arthur Compton wrote in 1956, the report "did not point to anything tangible that could affect the outcome of the war." Bush wrote Jewett that he was "rather disappointed" that the panel had not provided a clear-cut basis for making a decision for or against a full-scale effort and that he had "the same situation in my mind as before reading the report."[16]

Bush and Conant were not fully persuaded to support a fission project until they received the report of a committee of British scientists, the so-called Maud Committee, which argued convincingly that a fission bomb could be made within four years. Armed with this report, Bush arranged a meeting with FDR and Henry Wallace in October 1941 at which the president approved the preliminary project that Bush outlined for him. Roosevelt agreed to make a final evaluation of the project in six months' time and then decide whether there should be an all-out effort. In reality, however, the momentum generated in October 1941 largely predetermined the question Roosevelt answered affirmatively in March 1942.[17]

Bush and Conant agreed to commission a third NAS report in October 1941, one that would define technical procedures rather than engage in deliberation about feasibility. That debate had been closed. Compton and Lawrence were back on the team for the third report, and three additional academic scientists were also included: Robert S. Mulliken of the University of Chicago, W. K. Lewis, a chemical engineer, and George Kistiakowsky.

In a revealing letter to Jewett, Bush told his friend how the president's decision had clarified and circumscribed the role of the academic scientists who had challenged Bush's authority. Bush told Jewett that within hours of the October 9 meeting with Wallace and FDR, he "emphasized to Arthur Compton and his people [on the NAS committee] the fact that they are asked to report upon techniques, and that consideration of general policy has not been turned over to them as a subject." FDR's decision may have vindicated the arguments of Lawrence and Karl Compton, but it also ratified the dominance of Bush and his administrative approach. Bush had also reorganized his agency, subsuming the NDRC into the new Office of Scientific Research and Development (OSRD), which was created in July 1941. Bush and the OSRD was now the sole, undisputed representative of American science in the federal government.[18]

Bush's language to Jewett conveyed both relief and triumph. He happily passed the terrible responsibility of devising America's nuclear weapons policy up to the president, but he was also triumphant over his ability to withstand attacks from the academic stars of the profession: "Much of the difficulty in the past," he wrote to Jewett, "has been due to the fact that Ernest Lawrence in particular had strong ideas in regard to policy, and talked about them generally. . . . I cannot . . . bring him into the discussions, as I am not authorized by the president to do so. . . . I think he [Lawrence] now understands this, and I am sure that Arthur Compton does, and I think our difficulties in this regard are over."[19]

Bush's "conversion" and his method of resolving this momentous problem are revealing. He resisted the A-bomb project tenaciously at first because it seemed like a fantastic, impractical venture that might impugn the standing of scientists in the government. Since he lacked the knowledge needed to make this decision himself, he had to rely on others. Those he looked to with greatest trust were corporate scientists like his friend Frank Jewett. But they too were unable to provide a thorough analysis and explanation of whether the project was feasible. It was only when another government organization, from Britain, provided such an affirmative explanation that Bush and Conant were convinced. Bush would not rely solely on the arguments of his fellow American scientists to urge his government to take a decisive and possibly futile course of action. He was determined to remain "in harness" even if that harness was British rather than American. Most significant, he blocked his scientific colleagues from moving into policymaking positions. As the administrative head of the project, Bush would be the link or the ambassador between the scientists and the real policymakers in the White House and the Pentagon. He thus established a precedent of scientists acting as uncontentious collaborators.

Jewett remained skeptical. He worried about the government's influence in his profession and the potential damage that might result if the project failed. Fundamental problems remained, he pointed out. Was a chain reaction even possible? No one knew. The decision to proceed was based largely on assumptions, however theoretically plausible they might have seemed. There were also potential diplomatic problems: the British and the Americans offered very different estimates of the costs of the project. Might they also have very different ideas about its purpose? Jewett was particularly worried about someday having to go before an "appropriating body of laymen" to explain the decision to go forward. If that day ever came, he hoped better justifications for the decision could be found beside those on which they were currently operating. The whole problem reminded him of two useful bits of philosophy from his own experience: "One, 'if you can do it in a test tube you may be able to do it on an industrial scale, but if you can not do it in a test tube, you can not do it on an industrial scale'; and 'no matter how thin you slice it, it is still baloney.'"[20]

Bush responded with an explanation that suggests he was perceiving the emergence of a new standard in state science as opposed to the more cautious spirit of corporate science: "In time of peace," he wrote, "the things we are talking about would be almost absurd." But the emergency of a wartime race with the Germans changed the terms under which American scientists worked. Greater risks were justifiable.[21] But Bush still felt more comfortable with a protective governmental layer over him, a layer which he described in the language of business. He told FDR that "I would feel more comfortable if I shared responsibility, if I had something like a board of directors." Accordingly, the Top Policy Group was formed, consisting of Henry Stimson, George Marshall, Wallace, Conant, and Bush.[22]

Bush and Conant constructed their new partnership with the military on the basis of a practical, cooperative professional sensibility, one distinguished from either the supposed "salesmanship" of corporate engineering or the supposed "impracticality" of academic science. Throughout the project invidious distinctions would be made by the project leaders between academic and corporate scientists. One was constructive and valuable if one acted with the "realistic" and "hard-boiled" attitude of the corporate scientist. One was being "academic" if one clung to impractical or unrealistic ideas. In one instance the "academic" epithet was applied to a corporate scientist, Eger V. Murphree. Murphree, who was the head of Standard Oil's research laboratory, had been assigned responsibility for trying to separate U-235 through a centrifugal process. The process showed meager results, and a consensus began to build in favor of abandoning that approach.

Murphree, however, argued that there were good reasons for continuing. If the other methods failed, the centrifugal approach could still become a viable process if further research were carried out. Bush considered that argument illogical. If the other methods failed, the paltry centrifugal method would be of very little use; alternative methods would have to be chosen instead. Bush told Conant of his frustrating attempts to get Murphree to agree. The Standard Oil man, Bush complained, "has at no time met my point." Conant agreed, saying, "For an industrial man Murphree is the damndest 'professor' I ever met. He absolutely refuses to meet the issue."[23]

This standard of hardheaded practicality was also used to disqualify Albert Einstein from participating in the project. Bush toyed briefly with the idea of using the world's most famous scientist to perform some mathematical calculations regarding the rate at which gases moved through a porous barrier. Bush tried to get Einstein to come up with a calculation without telling him all of the factors involved in the problem or indeed the ultimate purpose of the calculations, which related to the separation of U-235 by a gaseous diffusion method. Given the limitations Bush imposed, Einstein's calculation proved not very useful. But Bush and Harold Urey agreed that they would not tell Einstein any more than they already had, and they gave up any further efforts to draw on his expertise. Bush, in fact, doubted Einstein's discretion, and Urey doubted whether the author of relativity theory had the mental attributes necessary to engage in the "hard labor" of carrying out a "long, difficult detailed calculation regarding" a concrete problem that was very different from the theoretical problems on which Einstein usually worked.[24]

Bush and Conant were not the only scientific administrators who were eager to show the compatibility between their profession and the military. Karl Compton addressed this issue in a 1942 book, *Scientists Face the World of 1942*. Misconceptions and problems had to be overcome on both sides of this new partnership. Many scientists in the OSRD, Compton wrote, "doubt whether the armed services give adequate recognition to the significance of the results which are being obtained and show as much enthusiasm as could be wished about putting the results into production and use." Some were "more imaginative than others and quick to grasp the significance of new developments" while others were "inherently conservative and find it difficult to accept the tempo of a wartime emergency." Some officers regarded scientists as "promoters who have some device or service to sell, rather than . . . as colleagues who are trying to assist them in accomplishing their purposes."[25]

Bush was determined to break the subordinate and demeaning position of scientists vis-à-vis the military, and once again he made his case by citing the examples of American business. In January 1942 he explained to Harvey Bundy: "Good practice, in the war effort, as in business, requires a reasonable independence of the organization of primary research from that of current development and production . . . with close liaison between." The scientists had to be able to do more than "simply attempt to supply the explicit devices called for from day to day by officers of the armed services." Such officers lacked the scientific and technological vision to perceive the types of weapons that would be possible and useful in the war. For scientists to provide that expertise, they had to be where the action was—in the field. He even suggested the creation of a "scientific corps."[26]

Bush himself clashed repeatedly with some of the highest-ranking members of the military over issues such as draft deferments for scientists and the military's slowness in deploying new weapons created by the OSRD. He left a vivid and scathing account in his memoir of his long-running agon with Admiral Ernest King over radar, for example. For their part, some military people were certain that Bush sought more than just greater recognition for scientific and technical expertise. Admiral Julius A. Furer complained once after a luncheon with Bush that the latter gave "the impression as he has many times before that he is very much put out over the fact that he is not called in by King to discuss the grand strategy of the Navy."[27]

These tensions were not experienced just in Washington. The tribulations of the Metallurgical Laboratory (Met Lab) at the University of Chicago have gone down in Manhattan Project lore as the most vivid instances of the tensions between the military, the academic scientists, and the corporate scientists from Du Pont. It was there that the titanic battles between General Leslie Groves and Szilard were waged. But analyzing Arthur Compton's management of the project at Chicago allows us to see the maturing professional and political identity of the corporate and academic scientists who were becoming "state scientists."

Compton's story is significant because he experienced in his own mind the tensions he was trying to smooth out between others. He was an academic scientist who, at crucial moments in the history of the program, resisted capitulating to the policies of the corporate scientists from Du Pont. But he was by no means an implacable foe of business participation in the project. Before the war he worked for fifteen years as a consultant for GE, and he also worked briefly for Westinghouse.

Like Bush, Arthur Compton at Chicago saw his own role in the project as that of a mediator between competing interests and objectives. In his 1956 book *Atomic Quest* he used language similar to Bush's and described himself as an ambassador between frequently warring groups. Diplomatic tact was definitely in order at Chicago; the inherent uneasiness between the academic and industrial scientists was intensified by the feeling among the academicians that they had a special responsibility for the development of the bomb since it had sprung exclusively from their community.[28] That tension was nowhere more evident than in late 1942, when problems were discovered about the feasibility of using plutonium. It was a major crisis in the project, and in the mind of Conant at least, it was another example of a conflict between the ideas and the attitudes of the academic scientists and those of the engineers of Du Pont. Compton found himself staunchly on the academic side of that battle line.

Impurities in plutonium interfered with the fission process by absorbing neutrons, thus cutting off the chain reaction. Just how pure did a plutonium sample need to be to be fully fissionable and thus capable of being used in a bomb? Up until November 1942, it was assumed by Lawrence, Compton, and Oppenheimer that as much as 3 percent impurities could exist in a critical pile of plutonium and still allow a chain reaction to take place. The British scientist (and discoverer of the neutron) James Chadwick had reached a radically different conclusion: impurities totaling more than 0.01 percent would kill the reaction.[29]

Conant was furious. He was particularly angry at the three American academic scientists who were aware of questions regarding impurity problems but had pushed those questions aside. Groves ordered an immediate review of the entire plutonium question by Du Pont scientists. They came to conclusions and made projections that were dramatically more conservative than Compton's estimates, to the point of being pessimistic about the entire enterprise. They estimated that there was only a 1 percent likelihood of producing enough plutonium to have a bomb ready by 1945. Compton had predicted that a bomb could be ready in 1944 and that one bomb per month would be produced in 1945. This discrepancy caused a crisis. The project had gone ahead only because the British, Lawrence, and Arthur Compton had argued that a plutonium bomb was feasible. Now the debate of the summer of 1941 seemed to be reopened. Conant lectured his subordinates: "The record, which some day will be gone over with a fine tooth comb, is of importance, not because of its effect on any of us, but because it will stand as to what American scientists can do under pressure. I should very much hate to have the record show that under the enthusiasm of the chase, American scientists lost their critical acumen and failed to be realistic and hard-boiled about the chance of success."[30]

In his own defense and on behalf of his Chicago colleagues, Compton wrote, "We feel that we . . . hold the primary responsibility to humanity and to science to see that our task is put in its proper place and is well done. . . . It is we who stand in the eyes of the public as the representatives of science." To both Conant and Compton this issue involved the public credibility of American science; they were not simply looking to the awful question of whether a miscalculation on someone's part could lead to the Germans winning the race for the bomb.[31]

In his 1956 memoir, Compton cast his discussion of these events in a more self-serving light. He ignored the mistaken estimates and instead described his insistent optimism in the face of the Du Pont executives' shocking skepticism. But in 1942 he was arguing that academic scientists working for a state at war were capable of making realistic assertions about what was scientifically and technically possible and making those decisions without applying the stricter, more cautious standards of traditional corporate practice. As Bush had written to Jewett, the actions of scientists in wartime, the gambles they were taking, would never have been considered prudent or practicable in peacetime. The standards had changed.[32]

The contemporary record of Compton's position in this "crisis" is not only more reliable than his memoir, it is also more provocative. In response to Conant's scolding, Compton argued that his staff at the Met Lab was as good as any group put forth by Du Pont. He pointed not only to the Nobel Prize winners on his staff but to the presence of corporate scientists. If Du Pont was skeptical and pessimistic, he argued, it should be dropped from the project. He offered to enlist the services of his former employers GE and Westinghouse.[33]

"It has been my lot," he wrote to Conant, "on previous occasions to differ radically with highly qualified scientific men on important questions." He was referring to his 1924 dispute with William Duane regarding the change of wavelength of scattered X-rays and his titanic struggle against Robert Millikan in 1932 over the electrical composition of cosmic rays. In both cases, Compton pointed out, neither man had accepted the evidence Compton presented. Their pride and egos forced them to engage in their own research, which took Duane two years to complete and Millikan five before they admitted their errors. Du Pont might take an equally long and unacceptably wasteful amount of time in an unsuccessful attempt to prove Arthur Compton wrong. That must not be allowed to happen.[34]

"A short investigation by 'practical men,'" Compton wrote, "can hardly fail to give a negative result, since our whole process falls far outside of industrial experience." This was the gist of Compton's argument: the criteria that corporate scientists applied to their own scientific problems had to be replaced by riskier standards. The wartime life-or-death race with the Germans made it irrelevant. The plutonium process had to work. If its feasibility seemed problematic in November 1942, the scientists had to find a way to make it feasible. He also argued that academic, not industrial, scientists should make fundamental policy decisions.[35]

The purity crisis prompted Conant to carry out a full-scale reappraisal of the project. He appointed a committee consisting of Warren Lewis (chair), Crawford Greenewalt, Thomas Gary, and Roger Williams of Du Pont. The Lewis committee made some criticisms but recommended the continuation of the pile project even before Compton claimed to have won the day by bedazzling the young Greenewalt with a ringside seat at Fermi's chain reaction experiment.[36] The plutonium project survived, but Du Pont's continuation was conditioned on its ability to organize the work as it, rather than the Chicago scientists, saw fit. The company organized the plutonium project in the same way it conducted its great commercial success, the production of nylon. The scientists at the Met Lab would continually chafe at the company's direction of the project, but they would never control the effort as they had wished.[37]

In the plutonium crisis, just as in the debate over whether to begin the fission project, scientists found themselves arguing that "normal" business criteria had to be superseded in the desperate race to create weapons before the adversary did so. This sense of working in special circumstances, of being insulated from the normal business tests of cost effectiveness, would come to constitute an increasingly important part of the professional identity of state scientists engaged in weapons research and development. This emerging scientific culture harshly chafed against a defining aspect of the corporate scientist's identity: the emphasis on practicality and good business sense.

The old problem that Jewett and Compton had faced resurfaced: scientists were innovators, but should they also be involved in the political management of their innovations? Wartime gave this problem a new, more dramatic twist because the innovations literally involved life and death. The administrative roles held by people such as Bush and Compton—"ambassadors" between their institutional sponsors and their professional colleagues—provided a mechanism for living with this contradiction. Scientists were doing their part by producing new technologies for the state. They would use state funds and resources, and a favored few like Bush and Conant could even act as advisers to the president. But like the corporate scientists, the wartime managers would cede a certain amount of political authority to the politicians and the generals.

Some of the consequences of that choice were evident during the Manhattan Project. Postwar problems of "security" and espionage were presaged in the experience of the Chicago lab, and Arthur Compton's own security problems are illustrative. He had briefly been a member of the Federation of Architects, Engineers, Chemists and Technicians, the leftist scientists' union to which Oppenheimer had also belonged in the late 1930s. This affiliation had given Compton trouble with Martin Dies and his investigative committee in 1941. Bush, in that instance, wrote the congressman and vouched for Compton's loyalty.[38]

But his membership drew renewed attention from Naval Intelligence, which wrote to Bush in February 1942, saying that it would clear Compton to work on matters of "pure science" but could not permit him to work with classified naval information of "strategic military importance." The military people also made an important distinction between "pure" and "applied" scientists. They clearly considered the work of the latter more directly relevant to security and thus more in need of policing than the work of the unworldly "pure" researcher.[39]

Bush, rather cleverly, exploited this false distinction to protect Compton. He wrote back to Richard Thayer, the chief of naval operations, and said that he concurred with their assessment of Compton. While he was undoubtedly a great scientist, Bush wrote, "I do agree with you that he is some what lacking in discretion, and while I feel he is using proper care at the present time, we have recently cautioned him in this regard and I feel that his continuance needs to be under conditions where both Dr. Conant and I will pay due attention to the matter for this reason." But after seemingly endorsing and mollifying the judgment of the security agents, Bush then proceeded to subvert their ruling by assuring Thayer that Compton was working on a new matter and "it is not necessary at any time to pass on military information" to him. Such a distinction would have warmed the heart of any lawyer. Clearly Compton was working on the kind of "new matter" that the navy people wanted to keep him away from. But Bush found the logical loophole: the fission matter was not yet a secret in the possession of the military. It still remained a theoretical project and was not yet a practical military reality.[40]

In a memo to Conant, Bush recounted the more explicit statements he had made to Thayer over the phone. He had told the navy man that Compton was undoubtedly loyal "but that he was somewhat naive and lacking in discretion and that you and I had been disturbed at times for this reason." Bush played pointedly to the prejudice against scientists that probably existed in Thayer's mind. He told him that there were a large number of scientists working on the project, "some of them of queer types and backgrounds, and that we [Bush and Conant] had thought of putting a couple of investigators at work." Bush then commented to Conant that the use of such investigators "would help us to strengthen our position."[41]

Compton remained skeptical and even at times openly resistant to the strict security attitudes adopted by the civilian as well as military leaders of the project. When Bush asked him to provide a list of all the scientists in the country who were known to have been informed of the project, Compton complied but added that if "rigid FBI tests" were applied in recruiting matters, most scientists would have been deemed ineligible. "Had we played safe in the sense of secrecy," Compton wrote, "we would have taken the greater risk of remaining ignorant of a powerful weapon which our enemies may use against us. . . . It is essential that we continue to be free to use our country's best talent on this problem, even at the risk of using men who are questioned on grounds of nationality and political partisanship. It is useless to attempt to develop this program without making use of the only brains our country has that are competent to handle it."[42]

The security concerns broached by the project leaders often deserved resistance as well as ridicule. Conant wrote a memo to Compton urging him to avoid hiring nonscientific laborers at the Met Lab whose parents were immigrants. Conant described the offspring of such parents as "near-aliens." "We should use men whose background is 'All-American,' for the routine jobs," said the Harvard president. Compton responded somewhat laconically, "There is, as you know, a large portion of the nation's labor personnel whose parents are foreign born," but he said he would try to meet Conant's request.[43]

The scientists under Compton's charge at Chicago continually fought against the leaders of the project, who fought back. When some of Harold Urey's criticisms reached the White House, Conant described Urey's complaints as "an extremely disloyal way for a member of the s-1 Committee to behave." Leo Szilard was the most famous gadfly in the Manhattan Project, and he essentially waged a one-man war against the administrative rule of Bush, Conant, and Groves.[44]

Even Compton found himself defending Bush and Conant against criticism as the project wore on. When Eugene Wigner threatened to quit over his dissatisfaction with Du Pont, Compton sharply objected. In contrast to his angry criticisms of Du Pont's role in the plutonium problem in November 1942, Compton now staunchly defended the company against Wigner's charge that Du Pont's real motivation for conducting the project was to gain control of the postwar atomic energy industry.[45] There was little basis for such a belief, and Compton pointed out that the facts of the matter actually contradicted it. Du Pont was very reluctant to take on the project, as Compton well knew; the company was not only fearful of being painted once again as a "merchant of death" but was skeptical about the success of the plutonium project. All patent rights had been assigned to the government to avoid any basis for the kind of claims that Wigner was leveling.[46]

Compton told Wigner that it was essential for the academic and corporate scientists to get along. "We now find ourselves with unusual power in our hands," he wrote. "If we cooperate with industry and show our adaptability to war conditions, with the objective of the common interest continually before us, we shall take an ever more responsible place in shaping the world." Dissent was especially inappropriate, Compton argued, since both the corporate and academic scientists were working at the behest of the nation's commander in chief. "Can a loyal citizen," he asked, "especially one in a key position, justify placing his judgment against the line of authority at such a time?"[47]

This deference toward military and corporate authority and concern about the postwar image of American science are also discernible in the debates among project participants over the question of whether the weapon should be used and how it might be controlled in the postwar world. Once again Bush's and Conant's positions indicated their willingness to manage and control this new technology and to steward it as a joint project between the scientists and the political leaders of the major powers. Along with other project leaders such as J. Robert Oppenheimer, they sharply rejected the more assertive position advocated by Leo Szilard and other Chicago scientists that called for a public discussion on the merits of using the bomb.

That position was expressed most eloquently in the so-called Franck Report. Like so many fresh and useful ideas in twentieth-century America, the report was inspired by an émigré scientist, Nobel Laureate James Franck. One of the most respected figures in the Met Lab in Chicago, Franck was the guiding spirit in creation of a remarkable document written by a group of Chicago scientists and given to Compton to bring to Washington. The report was drafted in June 1945, and it emphasized the unique political as well as military character of the new weapon, which demanded novel responses. "In the past," the report read, "science has often been able to provide new methods of protection against new weapons of aggression it made possible, but it cannot promise such efficient protection against the destructive use of nuclear power."[48]

This point flew in the face of the progressive, technological faith that had characterized the American discourse about science and technology for the preceding fifty years, when science was considered capable of curing any problems it created. Science was always one step ahead of its inherent dangers, and it would benignly move forward by itself. That mind-set was challenged by the Franck Report, which argued instead that people must create necessary safeguards, that unless scientists and politicians acted, the new weapon would proliferate and an unchecked escalation of the arms race would lead to disaster. The Russians were likely to develop their own bomb, the report pointed out, "within a few years." They had understood "the basic facts and implications of nuclear power" since 1940. Therefore the time to act was now.[49]

This too flew in the face of more recent conventional wisdom. Groves, who was impressed by the superiority of capitalist technology, believed the United States would enjoy a nuclear monopoly for as long as ten, perhaps even twenty years. The authors of the Franck Report anticipated this line of reasoning, which they correctly assumed would be used to argue that Americans need not fear competition with the Communists. Yes, the report responded, the United States would have many industrial and technological advantages in such a competition, but "all that these advantages can give us is the accumulation of a larger number of bigger and better atomic bombs."[50]

With regard to the immediate and pressing problem of whether to use the atomic bomb, the report was just as eloquent. It urged a demonstration before the Japanese and perhaps before the entire world. The demonstration would obviously not preclude later use, and an ultimatum against the Japanese could be issued which, if it did not bring surrender, could permit the evacuation of certain areas. The report also suggested that the approval of the United Nations and perhaps of the American public should be obtained before using the bomb. "This may sound fantastic," the authors acknowledged, "but in nuclear weapons we have something entirely new in order of magnitude of destructive power, and if we want to capitalize fully on the advantage their possession gives us, we must use new and imaginative methods."[51]

The report also addressed the argument about postwar congressional accountability, which suggests that the perceived need to justify the project financially by using the bomb may have been more widespread among the scientists and administrators of the project than most historical accounts have indicated. The authors questioned the assumption that "the American public will demand a return for their money." But, they pointed out, if the public was sufficiently educated about the horrors of nuclear warfare, it might develop the same revulsion toward those weapons that it showed toward the poison gases used in World War I. "It is not at all certain," they maintained, "that American public opinion, if it could be enlightened as to the effect of atomic explosives, would approve of our country being the first to introduce such an indiscriminate method of wholesale destruction of human life."[52]

The authors of the Franck Report stressed the need to consider the public's attitude about the moral and political effects of science and technology. Not only might the United States be morally besmirched in the eyes of the world, but the surprise use of the bomb might initiate an arms race in which the need for technical superiority would render public opinion meaningless. Science, they implied, could progress humanely only in accordance with the overall values of a society. It would not be enough simply to rely on the conservative, managerial stewardship of Bush, Conant, and other scientific statesmen.

The Franck Report was as much an attack on secrecy as on the bomb itself. Since the bomb could not remain a secret, the political question of its use and control also had to take place outside the cloak of secrecy. "The question of the use of the very first available atomic bombs in the Japanese war," the report's authors wrote, "should be weighed very carefully, not only by military authorities, but by the highest political leadership of this country." Later they reiterated that the decision "should not be left to military tacticians alone." They envisioned the creation of a postwar international agency that would exercise extrasovereign powers over the nuclear activity of nations. They closed their statement with a plea for the widest possible public discussion of this momentous issue. If the United States chose to have a demonstration, they argued, "it will then have the possibility of taking into account the public opinion of this country, and of the other nations before deciding whether these weapons should be used against Japan."[53]

Franck and Arthur Compton took their document to Washington, but they could not get an audience with the president or with Secretary of War Stimson. Franck ultimately met with Henry Wallace, who was no longer an insider with regard to the bomb project.[54] Compton himself hindered the effort to bring the report to Stimson's attention when he provided a cover letter which noted that it "did not mention the probable net saving of lives" or that the use of the bomb would show how essential it would be to avoid future wars.[55]

The project leaders themselves were now as responsible as the political and military leaders for accelerating the momentum toward unannounced use. Arthur Compton, along with Oppenheimer, Lawrence, and Fermi, was a member of the Scientific Advisory Panel to the Interim Committee, which Stimson had set up to advise him on matters of "control, organization, legislation, and publicity" attendant to the use of the bomb. On May 31, 1945, that committee held its historic meeting to plan the bombing, and it was there, by accident, that the question of a warning to the Japanese was first broached by Ernest Lawrence in an informal discussion during a lunchtime recess. Stimson directed the Scientific Panel to submit a report on the question.[56]

The panel traveled to Los Alamos to canvass the opinion of the scientists there. Oppenheimer did his best to shape the report toward a recommendation in favor of unannounced use. According to Compton, Lawrence remained the most determined to find an alternative to a surprise bombing. There were considerable differences of opinion among the scientists at the desert laboratory, but the report that Oppenheimer submitted to Stimson glossed over this disagreement: "The opinions of our scientific colleagues on the initial use of these weapons are not unanimous," Oppenheimer wrote on behalf of the panel. They could "propose no technical demonstration likely to bring an end to the war; we see no alternative to direct military use."[57]

The report seemed to emphasize its disagreement with the ideas set forth in the Franck Report, and in the language drafted by Robert Oppenheimer we can hear a continuation of the debate first initiated in 1940 and 1941 over the extent to which scientists should aggressively try to shape and lead policymaking. "With regard to [the] general aspects of the use of atomic energy, it is clear that we, as scientific men, have no proprietary rights. It is true that we are among the few citizens who have had occasion to give thoughtful consideration to these problems during the past few years. We have, however, no claim to special competence in solving the political, social, and military problems which are represented by the advent of atomic power." His "unofficial" position was even more explicit: when he refused to let a petition of Szilard's circulate, Oppenheimer argued, according to Teller's recollection, that "scientists had no right to use their prestige to try to influence political decisions."[58]

Oppenheimer, like Compton, was not eager to assert any special role for American scientists in the councils of the state. His view on the question of use seems consistent with the attitude Compton displayed regarding relations with Du Pont: the scientists would prove themselves worthy partners in the great enterprise that was under way, but they would not try to establish themselves as a group with a unique and independent role to play in government. They would be cooperative partners with the powerful business and political organizations in which they worked, nothing more.

The arguments over the proper relationship between scientists and the state continued into the early postwar years. Jewett and Bush, who shared a strong skepticism about the fission bomb in 1939 and 1940, found themselves divided in 1945 over the question of the kind of research agency that should exist after the war. Their differences suggest the diverging paths along which state science was developing. Jewett and Bush both agreed that a permanent research establishment had to be created after the war. Both also felt strongly that a lack of sufficient spending on research had weakened

American military preparedness and that such a lapse must never be allowed to occur again. But for the scientists, more research money was not an end in itself but a means toward the larger goal of enhancing their institutional and cultural status.[59] There was sharp disagreement over how to achieve that goal.

The various statements Bush and Jewett made concerning the question of postwar research agencies reveal important differences in their thinking about the proper role and even the proper institutional location of state scientific work. Bush wanted to create research organizations run by scientists in collaboration with the military. Jewett hoped vainly that he could get the Congress to strengthen the influence of the National Academy over military research and development.

Appearing before Congressman Clifton A. Woodrum's Select Committee on Postwar Military Policy in January 1945, Jewett testified against the creation of the kind of independent research board for national security that Bush proposed. Under Bush's plan, an equal number of people from the army, navy, and NAS would formulate new research projects for the services. "The Army and the Navy have a statutory responsibility for the defense of the nation," Jewett said, "which they cannot transfer to others." In May 1945, he appeared before Congressman Andrew May's House Military Affairs Committee, which was also considering whether to create a "permanent program of scientific research in the interest of national security." He told the committee that he was "opposed to setting up more agencies, on general principles," and he believed that he spoke for many scientists when he cautioned against the creation of a permanent government-controlled research institution.[60] Even those scientists already working in the OSRD, he pointed out, found the restrictive conditions of military research "repugnant." The scientists wanted to get back to "creative peacetime work." The intrusion of scientific experts into the traditional institutions of government also disturbed him.[61]

The NAS, Jewett told Woodrum's committee, was the "only organization in the United States which ties together all that we have of fundamental and applied science." But more important, he believed, it would be the only organization capable of carrying out postwar research with an appropriate regard for the public good. The spirit of the NAS, as expressed in its founding legislation, "imposed the most powerful obligation for disinterested service that can be applied to me," which was "the obligation to serve the nation without hope of reward."[62]

Jewett had made a similar argument three years earlier before Senator Harley Kilgore's subcommittee, which was investigating a proposed bill to establish a federal office of technological mobilization with broad powers over the work of American scientists. "The strongest authority that you can give to anybody is the authority of distinction without power, a thing that Mr. [Elihu] Root, who was very much interested in the Academy, always harped on. If you want to get things done, don't give people police authority to do it, make them so distinguished that nobody dares run counter to their desire." Once again, Jewett found himself very much out of tune with the emerging ideology not only within state science but within the state itself.[63]

Bush's views on the state of postwar science were guided largely by his historical memory of what happened after World War I: the brief connections between scientists, industry, and government forged during that conflict quickly dissolved. The political leaders of the country lost interest in the problem of military preparedness, and the inherent tendency of the services to resist innovation took firm hold.

"We must not go back to either the organization or the philosophy which prevailed with regard to scientific research on military matters in the years between 1918 and 1939," Bush told the Woodrum Committee in January 1945. The military could avoid repeating those mistakes, he argued, by adopting the same concern for scientific research that American industry had acquired throughout the preceding decades.[64] "The Services have not yet learned—as industry was forced to learn a long time ago—that it is fatal to place a research organization under the production department." Scientists needed to achieve the same degree of respect and independence in the military that they had struggled to achieve in American industry in the 1910s, 1920s, and 1930s. "There should be some form of partnership between civilian scientists and the military," Bush continued, and he urged that military officers receive better technical and scientific educations; West Point, he noted, slighted science in its curriculum. The criteria for promotion should also include an evaluation of technical expertise and not simply battlefield experience.[65]

But while Bush wanted to make the scientists equal partners with the "production department," Jewett thought that the history of the corporate research laboratories argued for the institutional independence of the scientific researcher. "Practically every industrial organization," he told Charles Wilson's Committee on Post-War Research, "started out with the idea that it should be closely connected with the operating departments in the determination of what it should do. Through experience practically every one has come in some form or other to a form of independent set-up with proper liaison with the operating departments it is designed to serve."[66]

Bush was after something more profound than a proper working relationship. Just as he had done in his battles within the NRC in the 1930s and just as corporate scientists like Jewett had done during the battles to legitimize basic research in industry, Bush called for a reorientation of both the services and the scientists so that they might form, together, a new entity. He sought to change the standards of merit from purely martial values to include the values of his own profession. In short, he wanted to imbue the military with some of the standards and sensibilities that existed in the corporate and academic research labs of America.

These were novel proposals, bordering on the radical, when one considers the stubbornness and resistance to change that existed in the armed services and against which Bush had already battled. "Military tradition," he told the Woodrum Committee, "has in the past called for planning in terms only of existing weapons. . . . The failure to have at the top levels of the military organization trained scientists and military leaders who plan in terms of future weapons or weapons in process of evolution may be costly in terms of lives and battles." The services needed to make use of people "of the intellectual fiber and background to enable them to synthesize the two types of thought, military and scientific, into an integrated whole." There needed to be a commitment to creating a "professional partnership between scientists and military men," in which the scientists would enjoy the status and working conditions comparable to those accorded by the "better university and industrial organizations."[67]

Bush's vision won out over Jewett's, of course. Yet in spite of finding himself on the "losing" side of yet another professional shift, Jewett comes out of this historical episode far the more prescient of the two men. Bush had been so concerned with elevating scientists and engineers to an equivalent status with the leading professionals in the American state that he seemed scarcely able to perceive any dangers amid his triumph. Back in 1942, Bush had joined Jewett in opposing the Kilgore bill and said, "In time of war, I believe in a centralized control . . . if every effort is made in the process to preserve the independence of action of scientific and technical groups within their assigned spheres of action to the maximum possible extent." That centralized control was in danger of becoming a permanent fixture under the auspices of the military. In 1945, as the government contemplated the manner in which it would continue its sponsorship of scientific research for warfare, Bush seemed inattentive to the dangers of losing the independence he prized.[68] Jewett, in contrast, seemed highly aware of the dangers of adding new limbs to the leviathan of the state. "We all know (Congress best of all)," he had told the Woodrum Committee, "that agencies established by law are easy to create and devilishly hard to modify or eliminate. They tend to aggrandize themselves and take on vested interests which they strive to perpetuate."[69]

Bush's attitudes about the proper postwar role of scientists can also be seen in his hostile response to the activist scientists in the atomic control movement. He did not object to scientists expressing their opinions, but he was annoyed when scientists assumed they were the supreme authority on nuclear matters. He was particularly irked by what he believed to be an air of superiority and even arrogance in the actions of some newly famous Manhattan Project colleagues who became involved in the battles over atomic energy legislation.

As early as July 1944, Bush and Conant had participated in some of the preliminary drafting of what became the May-Johnson Bill for atomic energy legislation.[70] Bush, however, was never drawn to participate in the "scientists' movement." He cordially and respectfully declined overtures from the head of the Association of Los Alamos Scientists (ALAS) to join with the public advocates of control. Alice Kimball Smith speculates that the reason why older scientists like Bush and I. I. Rabi demurred from joining their younger colleagues in such groups was a matter of temperament and age. Rabi, who was in agreement with the goals and methods of the movement, told Smith, "They didn't need us. They were doing just fine, and we older men would have spoiled the show."[71]

But such avuncular detachment does not explain Bush's reticence. He was greatly disturbed by the spectacle of scientists offering themselves as political and not just technical authorities. He found himself criticized by some scientists for supporting the May-Johnson Bill, "but when I pin them down I find they will attack me for lack of something in the bill when it is actually in the bill and when I put it there." The inexperience of some scientists in the political realm was "not surprising" to Bush, "although the fact that they do not realize how much they have to learn makes me a little pessimistic on this score." He was particularly irked by the "didacticism" with which certain scientists could address members of Congress. "If they do not look out," Bush warned, "they will end up by becoming discredited."[72]

"Nothing has annoyed me more," he recalled in 1953 to his friend Don Price, "than the attitude of the scientist who attempts to speak with authority on matters that are far outside of his own field." He recalled being particularly angry one day on Capitol Hill, "when Langmuir and Urey preceded me on the stand and I was having a fit as they proceeded to talk to Senators in the same way they would talk to one of their classes or a group of subordinates while the subject matter was one [about] which the Senators themselves were professionals and the men before them were sheer amateurs."[73]

Bush was content to quietly carve out an institutional niche for state scientists while avoiding the highly public lobbying efforts of the activists. But by the late 1940s, that institutional niche was disintegrating. This was partly the result of the deterioration of Bush's relationship with President Harry Truman, who seemed to resent Bush for refusing to tell him about the atomic bomb project when Truman was vice-president. On an even pettier level, Truman resented Bush's failure to appear at an award ceremony at which the president was to honor the OSRD director.[74] Although the two continued to have cordial personal relations, the president shut Bush out completely from any involvement in setting scientific policy in 1945 and 1946. Bush was briefly brought into the White House to help draft proposals and a communiqué concerning the conference on atomic matters with Clement Attlee and Canadian leader MacKenzie King. But the chaotic and amateurish character of this conference only served to further diminish Bush's estimation of Truman.[75]

Bush's frustrating experience as head of the Joint Research and Development Board (JRDB) from 1946 to 1949 (renamed the Research and Development Board [RDB] in 1947) also indicates the way in which his institutional base was eroding. The JRDB should have been exactly the kind of independent advisory board Bush desired. The former OSRD head even tried to bring in some of the luminaries from the Manhattan Project to staff the JRDB's three-member subcommittee on atomic matters: Conant, Oppenheimer, and Du Pont's Crawford Greenewalt.[76]

Bush envisioned the JRDB as something like a "court of arbitration" for the military's new weapons projects. The board would decide which service would develop a particular project, and it would try to avoid duplication of effort. But it would neither initiate nor terminate projects. The board consisted of two representatives from each of the military departments, and it enjoyed none of the power and influence the OSRD did. The principal reason for this was the existence of competing organizations within each of the services—organizations that were devoted to the pursuit of scientific and technical research such as the Office of Naval Research, the Army Research Office, and the air force's Office of Scientific Research.[77]

Bush's deputy at the RDB, Lloyd Berkner, shared Bush's ambitions for coordinating military research, and he shared his boss's frustration when those hopes proved vain. He offered a House subcommittee an illuminating retrospective on this failure. Both the JRDB and the RDB had been set up with various committees that were staffed with scientists who possessed expertise in the committee's area—radar, rocketry, and so on. These committee members then met with the military personnel who also shared responsibility for those fields. The result, Berkner wrote, was that the military personnel would ask how their existing technologies could be perfected; they were uninterested in any ideas the scientists might have about new weapons.[78] The bureaucratic imperatives of the services—to improve existing equipment and to prevent their "rivals" from gaining any advantage—prevented the rational coordination that Bush hoped to achieve. It also prevented Bush from having an effective voice in the research decisions being made in the services.

"There is great turmoil here on the scientific front," Bush wrote Jewett in May 1946. There was "an enormous amount of money being poured in and little real coordination as between Army, Navy and Air Force." Bush hoped for the speedy creation of "a top-line body with real authority to bring this into line." He had broached the subject "to the Joint Chiefs very forcibly and after a number of rather lively sessions" he thought he was making progress. But he was deluding himself. His irrelevance in the Truman administration was further highlighted when the president selected John Steelman instead of Bush to write a wide-ranging report on government-sponsored science in 1947. Bush, in the late 1940s, was rapidly coming to resemble Jewett of the late 1930s: a man who believed the institutions in which he worked could bring about the political goals he wanted to achieve yet who could not see that the fundamental interests of those institutions were very much out of synch with his own personal political goals. The fundamental interests of the armed services were to increase their power relative to their "rivals" and to ensure that they each fought off any attempt at military unification. Raising the question of some unified and efficient mode of conducting scientific work for the services took Bush "up to my neck in the whole question of a merger, for one angle of the thing cannot be handled without bringing in the whole subject."[79]

There would be no help from the top of the Pentagon. The secretary of defense at that time, James Forrestal, was in the grip of a severe mental crisis. But even at his best, Forrestal had been disdainful of scientists and felt they had no place in high policymaking councils.[80] Forrestal's suicide in 1949 seems to have triggered a mental crisis for Bush as well. He could not help but observe the parallels between their two situations. Both had been serving in Washington in exhausting official capacities since 1940. Both had increasingly found their power diminished and their goals frustrated. Bush was also nearing the point of physical and emotional exhaustion in the late 1940s. He recalled to Don Price in 1953 that while at the RDB, he "was working under considerable pressure, due to Jim Forrestal's position and his approach to unification, and moreover the attempt to put this whole thing into the condition that I thought it ought to be in pretty nearly finished me personally."[81]

In October 1948, Bush was hospitalized for exhaustion. He had also been suffering from intense migraine headaches that, he feared, might be indicative of a brain tumor. His doctors, however, found no organic causes and they concluded that the headaches were psychosomatic. Once Bush quit his job and left Washington on a long vacation, the headaches disappeared. His friend James Conant warmly seconded Bush's plans to quit the RDB. In January 1949, when Bush seemed to be restored to his old self, Conant wrote, "I am delighted to hear from everyone that your progress has been so rapid. All you needed was to get out of the Pentagon."[82]

The collapse of Bush's influence and the problem of interservice rivalry were ominous indications of problems that would reach full flower in the 1950s. Bush and the other wartime administrators would be replaced by a new generation of scientific leaders who had a very different set of beliefs about the role scientists should play in American political culture. Ironically, many of the ideas of Bush and his colleagues would be used against them by that later generation. To understand the scientific politics of the 1950s, we must delve deeper into the ideological landscape that Bush and the wartime leaders helped create.

## CHAPTER 4

## MAKING THE CASE FOR A
## MANAGERIAL DEMOCRACY

Much of the wartime work of the leading administrators of American science was secret. They were little known to the public and received only the faintest and most perfunctory oversight from a compliant Congress. But once the war ended, Bush, Conant, and a new addition to the leading ranks of scientific administrators, David Lilienthal, energetically placed themselves in the public eye and offered the institutional and ideological patterns of their wartime work as a model for postwar American political culture.

All three argued that America benefited from the collaboration among professionals, business, and the government and that such cooperation should become a permanent feature of American democracy. In making this case they laid the groundwork for the expansion of what Brian Balogh has labeled the "proministrative state" and a permanent system of institutional and professional collaboration among universities, corporations, scientists, and the military. Bush and especially Conant were directly responsible for implementing a meritocratic culture in America, one rooted in the universities and the professions, that would stand as one of the most significant and lasting achievements of postwar American culture.

The postwar years seemed to represent the cultural, institutional, and political triumph of the professional and political values Bush and Conant had espoused since the 1930s. But paradoxically, their very triumph paved the way for the emergence of a more aggressive and militarized vision of the role of science in American political culture. The ideas of these three administrators would be used in very different ways by a subsequent generation of scientific leaders. Momentous historical events such as the creation of the Soviet atomic bomb, the Korean War, and the arguments over whether to build a hydrogen bomb provided the catalysts that led to the undoing of the power and influence of the wartime leaders. But those events lay ahead. In 1945 there seemed little cause for concern and much reason for optimism about the triumph of their ideological and professional agendas.

Bush's influence with the new Truman administration was eroding fast, but his role as the leader of the organization that administered the atomic bomb project made him an esteemed public figure. That esteem was rein-forced by a series of books published in the late 1940s that celebrated the achievements of the OSRD and made the case for a tighter, permanent part-nership between scientists, industry, the universities, and the government.

Bush's writing in the late 1940s represented an elaboration of the tradi-tional concepts of corporate science, joined with a vision of an active state supporting basic research in the universities. He argued that "advances in science, when put to practical use mean more jobs, higher wages, shorter hours, more abundant crops, more leisure for recreation, for study, for learning how to live without the deadening drudgery which has been the burden of the common man for ages past."[1]

This was the familiar, utopian strain of corporate rhetoric. But Bush combined these notions with a startling new set of ideas about the relations between science and the state and the role of state power itself. He argued that "science can be effective in the national welfare only as a member of a team, whether the conditions be peace or war." He wanted to bring about a "professional partnership between the officers in the Services and the civil-ian scientists" of America. At the same time he also stressed a particular concern of his and Conant's: the meritocratic cultivation of talent. The state had a part to play in developing scientific expertise, and it had to do so along meritocratic lines. Ability, he argued, not "the circumstance of family fortune," should determine who received a scientific education in America.[2]

The essence of Bush's argument was that the government needed to do these things because industry and the universities could not do them on their own. But their corporate effort would result in a nation that was prosperous, militarily strong, and wisely developing its native intellectual talent. He took the essential ideas of the corporate scientists—the impor-tance of professional cooperation and the centrality of basic research to national progress—and moved beyond the conservative fears of state power that had characterized Jewett, Millikan, and other scientific leaders in the 1920s and 1930s.

The articulation of those ideas in *Science—The Endless Frontier*, more than anything else, accounts for Bush's prominent position in the history of American science. That book would be endlessly cited by later advocates of government funding for science and just as frequently cited by scholars of American science. But Bush also wrote another book in 1949 that is almost totally ignored but is crucial to understanding how he understood the political implications of the new arrangements he was helping to bring into existence.

That book, *Modern Arms and Free Men*, was intended to calm the public's anxieties about nuclear weapons by emphasizing how the new organization of American science and the American state would provide stability and security. It was a brief on behalf of the growing militarization of American life and a restatement of the meaning of democracy in a society dominated by expertise and by large, interdependent institutions. "The principal point," he wrote, "is that the atomic bomb is for the immediate future a very important but by no means an absolute weapon."[3] The ability of American institutions to foster cooperation and stability would make it easy for the United States to add atomic weaponry to its arsenal without threatening its political system. Indeed, the experience of World War II had already proven that dangerous new weapons could be created in America without disrupting the country's institutions: "I watched a great democracy bend itself around this new development," Bush wrote. "We contested with generals and admirals, but the new weapons were produced and used, and we wound up friends. . . . We did a job that required fantastic secrecy, and yet we won and held the confidence and support of the military, the Congress, and the American people. We were a varied group with all sorts of backgrounds and prejudices, and yet we developed a team technique for pooling knowledge that worked."[4]

Bush offered these organizational successes to the public as the greatest social and political innovation of the war. These techniques retained the essence of democracy: freedom and government by the consent of the governed. But that consent was being expressed in very different ways from that of the world of the town meeting or the political party system. Under the new setup, he wrote, "every citizen, in a strange and subtle way, visualizes where we are and where he feels we are going, and from this is distilled, in a way we hardly understand, our national policy in every regard." The amorphous, apolitical force of "public opinion" was able to strike this proper social and political balance seemingly without any human agency, just as the ideal state of Bellamy's utopia came about through effortless evolution. "How has public opinion done all this? In discussion and criticism our people somehow sense out the big issues, of course. But public opinion does this principally because men understand men and, through some process that is still mysterious, select those to be trusted, which is the essence of the democratic process. From the whole seething fracas emerge a national attitude and policy which become the guide for all who manage affairs in the public interest."[5]

Now, this seems entirely inadequate. Bush almost single-handedly created state science through his own political maneuverings in the late 1930s and early 1940s. For someone who had participated in the secret, closed decision-making process regarding the creation and use of the atomic bomb to speak so reverently about the power of public opinion seems false and even shameless. The public and its opinions had nothing to do with any of the work Bush carried out during the war. "Public opinion" served the same function for Bush that the idea of "grassroots democracy" served in David Lilienthal's book *TVA: Democracy on the March*.[6] Both ideas—fantasies, really—seem designed to reassure the authors as well as their readers that the essence of democracy was being preserved amid institutional forms that were patently undemocratic. The "grassroots" played no more a role in the administration of the TVA than did public opinion in the decisions made between scientists and a militaristic state.

But it would be wrong to dismiss such ideas entirely. They do have something important to tell us, I believe, about the terms on which elites in this new political system understood the boundaries of their authority. They also tell us something about the public's attitude, if only in broad terms. The public clearly had no role in policymaking, but its deference functioned as a form of ratification by default. The use of such abstractions by people like Bush and Lilienthal suggests their sense that they could justify their authority only in terms of serving some public need or by adhering to some broadly sketched conception of the public will.

Bush, in *Modern Arms and Free Men*, continued an intellectual and ideological strategy begun by people such as Theodore Vail, the social science experts, and the proponents of corporate monopoly in the 1910s and 1920s. We will later see that Conant and Lilienthal acted in this tradition as well. They understood their authority as a kind of deal between elites and the public: let us enjoy the new, unprecedented amounts of power we seek, whether that was monopoly control of an industry, or influence on policy, or government control of resources, or access by scientists to vast amounts of public monies; in return, we will act as an unoffensive ruling elite; we will use that power to give you things that you want: consumer goods, cheap electricity, an economically enriched academic sector that would sustain the nation's prosperity and security and would foster a meritocratic culture through government-funded scholarships and grants.[7]

They justified their power, in other words, on *perceptions* of the public will. But these perceptions had some legitimacy. The public *did* want what was being offered. They did embrace the consumer goods produced by the science-based monopolies. The vast increases in undergraduate and graduate education in the postwar decades testified to the desire of the middle class to enjoy the professional opportunities that education and participation in the corporate system afforded.[8]

Apart from these gross expressions of public will, however, we cannot speak meaningfully of public participation in these years regarding political decisions. And it must be emphasized that the idea of a meritocracy itself cut two ways: it augured a greater degree of democracy in terms of access to educational and professional opportunity for segments of the American public. But meritocracy did not just mean an objective, merit-based standard for selecting and promoting intellectual and technical talent. It also meant blocking out political ideas—and political conflict—from the government's sponsorship of scientific talent. Hence Bush's well-documented campaign against Harley Kilgore and Henry Wallace and their liberal notions of using science and social science for some form of public control over scientific development.[9]

An unmistakably dangerous aggregation of uncontrolled and unchecked political power was being organized into the institutions of state science in the 1940s and 1950s. As we will see, many of those dangerous potentials were realized in the political battles of the 1950s. But it is crucial to understand how leaders like Bush, Conant, and Lilienthal perceived their role. They did not see themselves as imposing a new order against the public's will but rather as leaders who were reconstructing American institutions by aligning and accommodating their interests with their conception of the public will. And their conceptions always excluded radical political positions and fit comfortably alongside the existing structure of the political economy.

This attitude explains the political restraint and disinclination to engage in public conflict on the part of the wartime leaders of state science. Since they were institution builders and alliance builders, and since they wanted to be accepted by other elites, they politically effaced themselves, just as the corporate scientists had done in the 1930s. Restraint was part of the deal. Their political influence rested, in their minds at least, on their ability to act above, or apart from, the grubby world of conventional politics. They did not want the public to see them as acting on behalf of any special interest. They were managers and mediators. Dictating to the public, as Jewett tried to do, or to other elites, as Bush would later try to do, would only bring rejection.

Given this reluctance to engage in political combat, it is no surprise that Bush also reached back to the ideas of the late nineteenth-century social scientists and resorted to evolution to justify the new power of experts in American political culture. The rise of expertise and its use in the Cold War, he argued, were both products of the evolutionary process. The human evolutionary path, according to Bush, moved in two directions that were in conflict with each other and were perfectly embodied in the contest between the capitalist and the communist worlds; it was the "agelong contest between those who would build and those who would dominate." The Cold War simply crystallized that conflict "into a new form."[10] In a synopsis of American history that was as terse as his interpretation of evolution, Bush argued that ever since the days of "Hamilton and his Federalists," elite citizens had been willing to step forward and organize themselves into effective associations that were capable of exerting social and political leadership. These associations frequently appeared as "stabilizing" forces in "days of peril." That stabilizing force, "in its worst form," Bush argued, "is a conspiracy of the money power to control. In its best form it is an alliance of men of outstanding ability to produce equilibrium by informal concerted action, upholding one another's hands for the purpose, joining to resist the demagogue, and placing the safety of the structure above the secondary policies and programs on which they might contend vigorously." America had "outgrown the need for the tight internal alliance of men of wealth," but it "will never outgrow the need for an unselfish alliance among men who place the stability of the republic above their secondary differences."[11]

In the conclusion of *Modern Arms and Free Men*, Bush celebrated the rise of a new public-spirited profession in America. "This is an amorphous profession," he wrote, "composed of men of good will in public affairs. It has no artificial solidarity emphasized by titles or symbols; it is bound together only by the common determination to make democracy work. It can bring order and reason out of chaos in our military planning." This group was responsible for the American future: "They will determine how all the rest of us function and how strong we may be. If they are numerous, determined, and united in the common purpose of preserving our freedom in a hazardous world, if we have sense enough in our peril to elect and support them, the future is safe in their hands."[12]

The existence of such an elite of civic professionals meant that Americans did not need to fear the possibility of atomic war. "A great war would be terrible," Bush assured his readers, but "it would not utterly destroy. It need not destroy democracy, for the organization of free men tends to become refined under stress, whether the stress be hot or cold, and meets its greatest hazards when the times are soft."[13] The military and militarism, according to Bush, triggered the best and most "efficient" aspects of American democracy. He believed that America could be militarily strong while simultaneously being responsible to the country's broader needs as long as the military establishment was broadened to include men like him, who would advise it and offer it wise direction.

Just as Jewett vainly hoped that American business would carry out his conservative political notions, Bush just as vainly hoped for an association of disinterested professionals operating at the highest levels of the military to keep the country armed and strong but doing so in a responsible manner. Evidence was pouring in, even as early as 1947 during the vicious battles over the unification of the services within a single Department of Defense, that the military was not going to behave well or act admirably. Instead, it was going to be mindlessly petty and would pursue its own institutional interests with little regard for the public interest.

There was one important exception with regard to the political restraint of both Bush and his friend and colleague James Conant. They both energetically advocated a stronger military, and both sustained the militaristic Cold War consensus that ultimately eroded their authority in the state. Bush and Conant played leading roles in the advocacy group the Committee on the Present Danger, which called for greater military spending and the enactment of universal military service legislation.[14]

Bush, however, continued to view his actions as nonpartisan and disinterested. He saw himself trying to preserve a "balance" that was constantly threatened by politics and pressure groups. In an October 1945 letter to Conant, for example, he wrote that the Truman administration was a "labor government" and warned of the danger of labor gaining the upper hand over industry. This eventuality had to be avoided. But even if this frightening prospect did come to pass, Bush was still convinced that there was an important role for people like him to play. Government by pressure group might be inevitable, "but it is going to work only if there is a sufficient body of the citizenry that sees the point, preserves the balance, and maintains the government in a position above that of merely a tool or an adjunct to the group that happens to have the ascendancy at the moment." Bush was somehow able to convince himself that these views were centrist and disinterested. "There is nothing anti-labor in my point of view," he said, "although it would be promptly interpreted as such. I would have been just as completely anti-capital in the same sense if I had been arguing a generation earlier."[15]

This centrism was, in Bush's view, above politics. And this constituted the essence of Bush's paralyzing political philosophy. He envisioned an activist role for scientists within the government, yet he hoped scientists would not take an assertive role in the public sphere generally. "By all means," he said, in a 1946 speech at Arthur Compton's investiture as chancellor of American University, "let us have plenty of public utterances by scientists speaking as uniquely qualified citizens on scientific matters, and let them be as authoritative as may be." But should scientists somehow act collectively? Bush seemed uneasy over such a prospect. "It is very salutary that the professional groups formed a detached unbiased section of the population," he said. They served as a "balance wheel" in the "evolution of our political procedures."[16] A strong military was simply part of his conception of a balanced state.

The man Bush was honoring that night in St. Louis adopted a similar view of the role of science and scientists in American life. By the late 1940s and 1950s, Arthur Compton saw the work of the wartime scientists and, indeed, the work of corporate science as a continuation of an evolutionary pattern. In his 1956 memoir, Compton asked whether it was morally justifiable to have introduced nuclear weapons into the world, and he drew an analogy between the introduction of atomic bombs and the introduction of fluorescent lamps in the electrical industry. The new fluorescent lamp was cheaper, more efficient, and ran longer than the standard incandescent lamp. Compton was a GE consultant at the time it was introduced, and he realized that the new lamp posed a grave economic danger to the electrical giant, but the company developed it anyway. The language Compton used was replete with the images of technological and Darwinian determinism.

"In my view," he wrote, "developments such as these are beyond human control. They represent, as I see it, an aspect of evolution. . . . We are destined, it seems, to try out new things, new ideas, and new ways of living. Some of these ways are more successful than others. Here is the competition for survival. The cost of failure to make the most of our possibilities is that those who fail are simply replaced by others." He said something very similar in 1946: "The evolutionary laws of survival of the fittest applies to societies as well as to individuals," he wrote; and continued "evolution" depended on cooperation and the absence of social conflict.[17] He had confidence that the rational calculations of business would prevent the possessors of the new atomic weapons from using them: "The hard fact is that war, like business, reduces to a question of gain versus cost. . . . In an atomic age, the resort to war demands a very high price indeed." The way to ensure that peace would prevail was to establish "ideologies and organizational patterns" that led people to "wider cooperation."[18]

Like Bush, Compton's ideal society was one in which there was no political conflict. America was a culture full of diverse groups, Compton noted, and "the society of the atomic age cannot tolerate the development of antagonisms among such groups. Much more than in earlier times it becomes a condition of survival that we shall love our neighbors as ourselves with a love that finds its expression in service." If the new society of cooperative institutions was to work, people must agree on goals, and there must be no deviation from those goals. If society was to be orderly, he preached, "the choices of its citizens must be harmonious and not in conflict."[19]

James Conant was also trying to use the new prominence of science to advance certain political goals in the late 1940s. At the same time, Conant, like Bush, was also helping strengthen the militaristic consensus that was infecting American politics in these years.

Conant's actions in the Cold War must be understood as part of his ongoing agenda to use the culture of expertise, with its interdependent relations among professions, state, business, and academia, as an instrument for creating a more meritocratic and less class-conscious society. He had been pushing this cause from the earliest years of World War II. In a July 1942 *Atlantic* piece, Conant argued that "native talent developed by education is what the army requires for leaders." He wanted the government to choose its officers "purely on the basis of merit, without regard for their economic situation," and he also wanted it to finance "this group for whatever further education was required by the Army or the Navy."[20]

"Ability," he declared, "must be discovered and financed by the government so that the very best men will be available for the nation's needs." He wanted meritocracy elevated to a central war aim: "The very cause for which we fight, rests on the flexibility of our national life—it rests on our denial of the doctrine of hereditary privilege." America must continue to repudiate "the idea of a ruling class" and affirm "the ideal of a classless nation." If it did so, the reformed system would "both increase the effectiveness of our leadership in battle and demonstrate the reality of our American ideal." It would help win the war and at the same time "lessen those tensions between economic groupings which in the modern world are ever in danger of threatening a democracy's internal peace."[21]

The war, for Conant, rendered the existing categories of political thought anachronistic. In a 1943 *Atlantic* essay titled "Wanted American Radicals," Conant argued that "total war has automatically eliminated the conservative." Reform and even the transformation of postwar American society were inevitable. The Roosevelt-hating members of the business class who wanted to go back to Coolidge economics were not properly called conservatives but were instead reactionaries. Their opponents were not the "liberals" or the "Progressives" but the "radicals," that is, the people who were dealing with the unalterable reality of change in a complex world.[22]

The American "radicals" had to undertake "the work of redefining culture in both democratic and American terms." They had to continue to push the cause of meritocracy and be "lusty in wielding the axe against the root of inherited privilege." The American "radical" would be one who favored decentralization and local responsibility and who opposed excessive federal power. But he would be willing to use that power "in the interests of maintaining real freedom among the great masses of the population."[23]

Conant provides a fascinating example of political thinking that was new and yet also limiting. It echoed Croly, Lippmann, even John Dewey, with its pragmatic, democratic tone. Yet it also offered a vision of political life, much like Lippmann's, that limited the range of possible thought and action while simultaneously championing the cause of "flexibility" and greater opportunity. It replaced a political vocabulary and a political perspective that was, indeed, anachronistic with a new configuration that was excessively rigid and did not seem to allow for much variation of thought. The mentality of the American "radical" seemed fully formed. Were there not to be any differences of opinion over *how* these new powers of the state were to be used? Were there no grounds for disputing the character and the purpose of the new educational opportunities that were to be created? Meritocracy, in other words, for what?

Conant's actions in the early years of the Cold War suggest an answer. He saw the role of experts as being essentially a solipsistic one. Elites should uphold the authority of elites. For example, we know thanks to James Hershberg's excellent biography of Conant that he worked hard behind the scenes to rebut the growing chorus of condemnation against the atomic bombings. "You may dismiss," Conant wrote Harvey Bundy, "all this talk as representing only a small minority of the population, which I think it does. However, this type of sentimentalism, for I so regard it, is bound to have a great deal of influence on the next generation. The type of person who goes in to teaching, particularly school teaching, will be influenced a great deal by this type of argument."[24] Conant urged Karl Compton to write in support of the state scientists' decision to use the bomb, and the result was a December 1946 *Atlantic Monthly* essay titled "If the Atomic Bomb Had Not Been Used." But his more significant contribution to the elite defense of the bomb was his intense participation, along with McGeorge Bundy, in the drafting of Henry Stimson's 1947 *Harper's* magazine piece, "The Decision to Use the Atomic Bomb."[25]

By the late 1940s, as the Cold War consensus was infiltrating ever deeper into American political and cultural life, Conant clearly acceded to that consensus. He became increasingly tentative in his advocacy of the arms control initiatives he and Bush had so forcefully tried to bring to FDR's attention in 1944. And he bowed, as he had at Harvard in the 1930s, to state demands for loyalty oaths and for security investigations. He may have advocated a lusty application of the ax of meritocracy to inherited privilege, but he showed no such aggressiveness in the face of the militarism that was spreading over America's political institutions and civic culture.[26]

But Conant was more than just meekly acquiescent. He acted directly to further the Cold War ideology that led to precisely the secrecy and excess he decried. When scientists became the targets of the red-baiters in Congress, he did not merely refuse to help; he strengthened the government's case. When the House Un-American Activities Committee attacked Edward Condon, the new president of the National Academy of Sciences, Alfred Richards asked Conant if the academy should get involved and defend Condon. Conant wondered "whether the National Academy charging into this won't do more harm than good."[27]

In December 1948, Conant agreed to serve along with Oppenheimer and Oliver Buckley on an NAS panel directed to study the question of scientists' civil rights in the face of increasingly intensive and damaging "loyalty" investigations. Conant directed the panel toward the conclusion that "the government, in resolving doubts on these matters . . . must settle the case in favor of the government rather than the individual. If a shadow of doubt exists, the individual should be prevented from having access to confidential information."[28]

Conant also participated in the statement of the Educational Policies Commission (EPC) of June 1949 in which the commission agreed to the proposition that Communists should not be allowed to work as teachers. He tried to make a distinction between a policy against hiring and a policy of discharging Communists currently employed. "If you say they should be discharged, you are making an operational statement," Conant told a fellow EPC member. "You are ducking the issue, but you would want to duck it in a policy statement." Conant's entire career in the public realm seemed to be one of "ducking the issue" when his professed principles came into conflict with the prevailing tendency of the country's governing institutions.[29]

Conant's writings of the late 1940s and early 1950s, like Bush's 1949 book, served further to legitimize the Cold War consensus, while simultaneously playing down the notion that scientists had a prominent role to play in public affairs. At precisely the moment when the notion of an activist scientist was coming under attack from within the government, Conant was advancing a series of arguments designed to diminish the public's expectations about the roles scientists could play as public intellectuals and experts.

In two books—*Education in a Divided World* (1948) and *Science and*

*Common Sense* (1951)—Conant mounted a vigorous attack on the inflated and quite abstract hopes that the public had brought to bear on the "scientific method." "In the last twenty-five years," he complained in 1948, "indoctrination in the scientific method has been put forward with more and more insistence as one of the primary aims of modern education," and this worried him. "I frankly do not know what my friends and colleagues have in mind," for it seemed as if the proponents of the "slogan in question" proclaimed the scientific method "as a way of looking at life; at times it seems almost a panacea for social problems." To put the scientist on such a pedestal, Conant believed, was "quite erroneous."[30]

"With idolatry of science I must confess I have little sympathy," he wrote in 1951. Conant wanted the emphasis placed on the institutions of science rather than its methods. "The important fact that emerges from even a superficial study of the recent history of the experimental sciences," he wrote, "is the existence of an organization of individuals in close communication with each other." Because of this organization, "new ideas spread rapidly, discoveries breed more discoveries, and erroneous observations or illogical notions are on the whole soon corrected." There was "deep significance" to the existence of this organization, and most people failed to grasp it. "Indeed," he wrote, "a failure to appreciate how scientists pool their information and by so doing start a process of cross-fertilization in the realm of ideas has resulted in some strange proposals by politicians even in the United States."[31]

He wanted the public to distinguish sharply between the work of scientists as scientists and their actions as citizens in the larger public sphere. In the laboratory, the professional standards regarding the fair handling of evidence and data were so firmly in place that scientists dared not violate them. "Let him deviate from the rigorous role of impartial experimenter or observer at the risk of his professional reputation." There was nothing especially noble, in other words, about adhering to rational standards of logical thought and action. "Under these circumstances even an unstable personality may within the rigorous confines of a laboratory science become an impartial and judicial inquirer. But once he walks out of the door of the research institute—then he is as other men." And as mere citizens, Conant noted, "scientific investigators are statistically distributed over the whole spectrum of human folly and wisdom much as other men."[32]

He wanted to dispel the notion that the scientist offered some unique vision or approach to social problems. "Too many educators appear to underrate the amount of hard-headed thinking which has been done by practical men in the history of the human race," Conant wrote in 1948. "We must stress the significance of rational inquiry throughout our general education, but the identification of this type of inquiry with science confuses rather than clarifies the presentation." The important point to be made was that the hardheaded rationalism of the scientist was indistinguishable from the hardheaded rationalism of the lawyer, the soldier, the statesman, and all the other "practical men" in human history.[33]

"As a matter of honest pedagogy," Conant advised, "it seems to me that one of the first objectives in teaching the natural sciences is to show the frame of reference in which they now operate." That was the important point: to show the sciences and the scientists as only one branch of a larger social structure. "Rather than leave in the minds of the pupils the very dubious proposition that the methods of science are applicable to all manner of practical human affairs, we should show how legal methods of inquiry have been used in Anglo-Saxon countries. Likewise, we must study the rational methods of merchants, manufacturers, soldiers, and statesmen which were employed with considerable success for generations, long before any idolatry of the word 'science' came over the academic horizon."[34]

By way of illustration, Conant offered the readers of *Science and Common Sense* episodes in the history of science in which important new discoveries were made, and he was eager to show how difficult it was in some instances for the new idea to break through the encrusted prejudice of established knowledge. He also took pains to dwell on the failures of the discoverers, who often stumbled blindly around a new truth until they realized the errors they had been making in either their methodology or in their very conception of the problem. The procedures of science were always stronger than most of the ideas produced by that procedure. It was the procedures that always provided the instrument for advancing new questions, for pushing toward a deeper or different understanding of nature.

Conant was not simply interested in providing a more realistic public attitude about scientists; he also sought to impress the public with the integrity and the authority of the professional elite generally. By stressing the professional structure of science, he made a case for respecting and trusting the structure of elite institutions, regardless of whatever particular idea or policy that structure might produce at a given moment. An incorrect scientific theory might hold sway for a time, but the mechanisms of publication, criticism, and exchange and the admission of new members into the profession would bring about a better, more accurate theory in time.

Likewise, the new governing system of collaborating elites that was taking shape in America was more important than any particular policy they might produce. As long as those elites were effective and accessible they would foster a stable democratic culture, or at least the appearance of such a culture. And a stable democratic culture, in Conant's view, was not one in which social differences were overt or one in which political disputes and class conflict were openly handled by the nation's leading political and cultural institutions. Instead, it was one in which class differences were masked.

American institutions, he believed, echoing his wartime writings, had to obscure the class structure that existed in the United States and to blunt the possibility of conflict. In *Education in a Divided World* he wrote that there should be a "minimum of emphasis on class distinction," there should be "low visibility" of group differences.[35] As an educational leader, Conant wanted "to insure that our American idea of a classless society will stand firmly in opposition to the Soviet challenge," and education, Conant believed, was the forum in which America would succeed or fail in creating a classless society.

Bush and Conant blended together a powerful set of ideas: they tied the utopian strands of corporate science with its emphasis on progress, harmony, and abundance to newer notions of Cold War militarism. Scientific talent was essential to both the corporate and military sectors, and American scientists, they argued, had an important yet circumspect role to play in this new world. But Bush and Conant were fading from center stage even as they articulated this vision in the late 1940s. They were moving away from the center of political power, even if they did still enjoy considerable respect and notoriety. Their ideas and their institutional creations, however, lived on after they left Washington, and we need to examine how those ideas fared in the increasingly strident political atmosphere of the late 1940s and early 1950s.

Perhaps nothing more poignantly reflected the problems involved in trying to continue Bush's and Conant's approach to managing state science in the Cold War than the career of David Lilienthal. His experiences in the abortive arms control efforts of 1946 and his chairmanship of the Atomic Energy Commission (AEC) from 1947 to 1950 revealed the problems and the dangers of trying to achieve an apolitical reorientation of American political culture under the aegis of expert collaboration and state science.

Lilienthal hoped that his experience in the "new field" of atomic energy would allow him to replicate the professional cooperation and efficient, democratic organizational techniques that he had applied in his job with the Tennessee Valley Authority. He set forth the lessons of his TVA tenure in his 1944 book *TVA: Democracy on the March*, and his ideas were strikingly similar to those of Bush and Conant: he envisioned a new organizational paradigm for American society in which class conflict and the "old-fashioned" politics of battling interest groups would be replaced by a professional and cultural system in which social problems were dealt with and resolved through the rational and efficient application of expert knowledge.

In addition to such Lippmannesque "echoes," Lilienthal's book also reflected some of the ideas of Bellamy and the early corporate scientists. He argued that the purpose of adopting new organizational approaches to social problems was to ensure that individual citizens were able to enjoy the greatest degree of material comfort. The very idea of politics seemed to be banished in Lilienthal's philosophy in place of a bland, benign, and seamless philosophy of consumerism. It is worth examining Lilienthal's book in some detail to appreciate its place in the emerging ideological consensus to which Bush and Conant also contributed and to understand how the ideas in it shaped Lilienthal's conduct on atomic energy matters.

Above all else, Lilienthal believed, the TVA experience showed that "there is almost nothing, however fantastic, that (given competent organization) a team of engineers, scientists and administrators cannot do today." Those words can serve as a synopsis of the corporate scientific creed: it emphasized the connection between professional cooperation and social improvement, and it implied the primacy of experts and their knowledge over other forms of social action.[36]

For Lilienthal, the genius of the TVA was the way in which it recognized what he considered to be the altered character of American political culture. Expertise and organization had gained primacy in American political life because they were the most effective means of providing people what they wanted, which was material comfort rather than some political "truth." People were concerned about "how to get a job at a new kind of factory . . . at good pay; about a pleasant town where the kids can have bicycles; about electric lights and heated schools and churches and hospitals for the ill." The "dominant political fact of the generation," Lilienthal declared, was that people no longer looked upon poverty as inevitable. The TVA was in tune with this new spirit. It was an institution that moved beyond the existing ideological categories of the day, left versus right, private enterprise versus socialism. These were categories that had little relevance to the material and personal concerns of the ordinary citizen. "People," he wrote, "seem less and less beguiled by abstractions and vague eloquence; they seem no longer greatly moved and lifted by abstractions." People's needs were decisively individualistic: "A man wants to feel that he is important," Lilienthal noted. "This hankering to be an individual is probably greater today than ever before." The TVA could satisfy those longings. Resource development and its manifold effects could provide the satisfaction lacking in modern society. "Men would not only have more things; they would be stronger and happier men."[37]

Most important of all, TVA affirmatively answered the question that Lilienthal regarded as the most important of the age: could "business and the common weal both be served?" The TVA sought to meld public and private interests in "voluntary and non-coercive ways." It was not intent on pursuing "negative regulation," what Lilienthal dismissively referred to as "the everlasting 'don't do it again.'" It sought "to make affirmative action in the public interest both feasible and appealing to private industry."[38] The TVA was able to do this, he argued, because of its fresh and sophisticated perception of the needs of the public, its understanding of the role of the government, and its proper coordination of specialized expertise. The actions of the TVA all served to enhance the concrete, material needs of individuals. Once the state and business began cooperating to satisfy those needs, then social peace and harmony would prevail.

The definition of "public interest" Lilienthal employed differed, he hastened to point out, from that used by other liberals. He did not mean "the people" in their "institutional roles as wage earners or investors or voters or consumers." He meant simply "people as human beings," people as individuals. All of the benefits of the TVA that he described in his book were ones that redounded to the personal enrichment of individual men and women. When he spoke of "creating wealth" and of raising "the income level of the people," he was speaking on a distinctly microeconomic level.[39]

His understanding of the role of the state also differed from the traditional understanding of liberals. He did not see the government as merely an "umpire" that intervened to settle disputes between warring interests. But neither did he see the government as a "combatant" and an antagonist to business or labor. "The technical services of the government have a *job* to do—for business, for labor, for consumers, for the sustained productivity of us all."[40]

Few people realized, Lilienthal observed, how greatly social conflicts could be reduced through the effective cooperation of experts. The new state authority was so successful in large part because—like Croly's enlarged individual—it made all problems one problem: the effective operation of the authority's business. "In the way TVA goes about its responsibilities,"

Lilienthal wrote, "there are no 'jurisdictional' lines, no excluding of the chemical engineer, say, because this is a 'farm' problem, or of the business-man or the inventor because soil erosion is an 'agricultural' problem, or of a county or state expert because agriculture is a 'national' question." TVA experts cooperated with each other because "they work under a single management. That helps to unify their efforts and their thinking."[41]

Since the TVA's avowed purpose was to emulate the unity of nature, it could homogenize the competing interests in the region into this larger, transcendent goal. "It was not," Lilienthal notes, "conceded that at the hour of Creation the Lord had divided and classified natural resources to con-form to the organization chart of the federal government. . . . What God had made one, man was to develop as one." Such a philosophy made the role of politics in the authority's operations ambiguous at best. Lilienthal went to great pains to show that there was nothing political about the procedures or the operations of the authority and that political consider-ations played no part in matters of hiring or development. The Congress had expressly forbidden the use of the authority as a patronage mill. The fundamental work of the TVA may have transcended politics and reflected the will of the Lord, but it was inevitably bound up in a larger complex of political considerations. "A river has no politics," Lilienthal noted. "But the question of whether a river *should* be developed is a political question, and hence a proper subject of politics."[42]

Lilienthal was savvy enough to realize that there were many species of political action, and he was no great admirer of the antidemocratic, good-government elitists who often rallied to the causes of regional development, public authorities, and technocratic leadership. He even seemed to prefer the outright corruption of the machine hacks and the patronage dispensers to the righteously disguised self-interest of the reformers who practiced "a kind of Phi Beta Kappa version of Tammany Hall." But Lilienthal's writings reflected little recognition of the existence of social classes or of politi-cally distinct groups with incompatible political interests. Even when he spoke of "businessmen" and "organized labor," he used the terms merely to characterize the cooperating participants in the authority's endeavors. "The 'sights' of businessmen had widened," he argued, as a result of the successful operation of the TVA over its first decade. "Most businessmen," he wrote, in language that was reminiscent of Walter Lippmann's celebration of the "new" businessman, "have accepted as their own the TVA idea of region-building quite as wholeheartedly and understandingly as have farmers and industrial workers."[43]

This view was also reminiscent of Bellamy. Lilienthal spoke frequently throughout the book of the seamless web that the TVA supposedly represented. But unlike Bellamy—and in the spirit of Croly—he offered a concrete prescription for how this unity was to be brought about: it would be through the application of "modern management" principles through agencies of the federal government. Experts acting through the federal government were the agents that would bring about the fusion of public and private interests. It was Bellamy's cooperative utopia, mixed with Lippmann's celebration of the scientific expert and Croly's activist state.

The public's role was clearly not Lilienthal's principal interest when telling the story of the TVA, despite his paeans to grassroots democracy. The public functioned chiefly as the recipient of the expert's success (whereas the black public was nonexistent, as the TVA conformed itself to the Jim Crow exclusions practiced by the white population of the valley). The rising standard of living provided evidence of the scientist's, engineer's, and administrator's effectiveness. The experts and their organizational techniques were the real story of the TVA, for it was their effectiveness and cooperation that provided a "seamless web" of professional collaboration. That cooperation proved it was possible to unify the proliferating number of technical specializations in the modern world, a proliferation that Lilienthal regarded as one of the "spiritually disintegrating forces" of modern life and one of the "central problems" of the age.[44]

Lilienthal's TVA book advanced a view of scientists and engineers as the craftsmen of the "seamless web" of an ideal American society, a society geared toward the production of consumer comforts and not toward the clarification of political and economic rights of distinct and divergent groups. Lilienthal also shared with Conant, as well as with Bush, little faith in the "scientific mind" per se. His experts were of value only to the extent that they were participants in corporate institutions like the TVA.

When Lilienthal was appointed to head the State Department's Board of Consultants regarding the international control of atomic energy, he considered it "a natural and proper part of my work as chairman of the TVA" and "part of my education as a leader in seeking to bridge the desperate and continuing crisis of modern science and the human spirit. This is a major issue of statecraft." Throughout the weeks that the Acheson-Lilienthal panel met, Lilienthal's cooperative vision seemed to be materializing. Bush, upon the completion of the report, commented to his aide Carroll Wilson that he "couldn't have told which were scientists, which were industrial men, which the administrator of a public enterprise."[45] The report itself provides eloquent testimony to the extent to which men like Oppenheimer, Lilienthal, and Bush believed their institutional philosophies could be applied to the gravest problems of diplomacy. Reading the report, it is hard not to feel admiration for their ambitions, even from a distance of half a century and with full knowledge of how completely their vision was rejected.

The report noted, first, that its authors were drawn from a range of professional fields. This diversity, they stressed, was essential in handling the problem of nuclear weapons. "We are not dealing simply with a military or scientific problem, but with a problem in statecraft and the ways of the human spirit."[46] The report tried to do something that was all too rarely attempted in the long history of the Cold War: it tried to explain to the public what to expect from scientific innovation, and it offered an assuring vision of scientists playing a constructive role as technical interpreters for the public as well as for governments.

The plan's strategy of controlling uranium provided the key to its technical success, and Oppenheimer's words sought to strengthen the public's confidence in that approach, in spite of wildly inaccurate and often hysterical public comment about potential nuclear dangers. People in the newspapers and on radio, for example, foolishly speculated that any kind of matter, even clay or wood, could be converted into atomic energy, with great destructive effects. Uranium was the essential ingredient of any nuclear weapon, he promised, and the public had to accept the fact that there was no vast terra incognita that might bring some new nuclear horror upon the world. The calm, intelligent language used in the report was notable for its divergence from the fairy-tale quality of so much American writing about scientific change, whether benign or malignant.

"Novelty will of course appear in scientific discoveries, but it will appear for the most part not as a negation of present knowledge, but as the result of new types of physical experience made possible by new methods of physical exploration, and in turn requiring new modes of description." Those new discoveries might "have something to do with the basic knowledge involved in [the] release of atomic energy, but there is no basis for believing this, and the chances are against it."[47]

Responsible political leaders who were skeptical about the report's scientific claims "should test for themselves the correctness of our conclusions." Such testing would "require an examination of difficult and complicated technical facts, but we are confident that the process is one which other laymen, with the appropriate help of experts, can readily repeat." The report, in short, offered a realistic and appealing vision of how scientific expertise could be brought to bear on nuclear energy and how that expertise could serve to calm and educate a nervous public in the early years of the atomic age. The internationalization of atomic energy development, with its potentially lucrative civilian applications, might even prove to make the plan highly appealing. Perhaps people would be drawn to support it, if not actually participate in its operation, out of the desire to be " 'in on the ground floor' of a growing enterprise."[48]

The public received the report with great enthusiasm and commended its authors. Lilienthal seemed on his way to a grand elaboration of the ideological principles he had formulated in the TVA: progress through institutional cooperation and scientifically minded management of resources. But he and his collaborators had not figured on the intrusion of politics in the person of Bernard Baruch. On March 17, 1946, Truman, at the recommendation of his secretary of state James Byrnes, announced that Baruch, aged seventy-seven, and in what he himself charitably described as the "late afternoon" of his life, would be the special U.S. representative to present the American arms control proposal to the United Nations. Lilienthal was furious at the appointment and could barely hide his contempt for the venerable presidential adviser; he recorded in his diary that he felt "quite sick" over it. "We need," he wrote angrily, "a man who is young, vigorous, not vain, and whom the Russians would feel isn't simply out to put them in a hole, not really caring about international cooperation. Baruch has none of these qualifications."[49]

Baruch did indeed take the American proposal in a hard-line direction that Lilienthal and his colleagues on the committee abhorred. Bush refused to work with Baruch, and Oppenheimer despaired of any diplomatic success. Lilienthal deliberated about whether he himself should cooperate with Baruch. "If we nurse [Baruch and his aides] along to try to affect their decisions, we get nowhere and are blanketed; if we don't, we'll have to try to make a public ruckus, which isn't good either." He watched silently as the Baruch proposal failed to win approval in the UN.[50]

This relatively small episode was something of an omen of what would follow when Lilienthal became head of the new Atomic Energy Commission: initial high hopes would be replaced by political backbiting and conflict, leaving Lilienthal at a loss over how to implement his cooperative ideals in a decidedly uncooperative world. Almost immediately he and his scientific colleagues on the AEC's General Advisory Committee (GAC) became disillusioned about the possibility of realizing the goals and the values he had expressed in his TVA book. Instead of demonstrating the wonderful benefits that flowed from the cooperation of professionals and the state, the experience of the AEC in its first three years only showed the reckless way in which the Congress and the military would try to seize the issue of atomic energy as an instrument for pursuing their own institutional and political interests.

In the case of the politicians in Congress, that meant seizing on the issues of "security" and "loyalty." In the case of the military, the existence of the nominally civilian AEC represented a threat to its interest in controlling the nuclear arsenal. Both groups tried relentlessly to undermine the effectiveness of the new agency, and by the time Lilienthal resigned in early 1950 he had become deeply bitter and cynical about the usefulness of the commission or even about the ability of professionals to act effectively in the government. The congressional members who involved themselves in atomic energy matters could see the new technology only as an opportunity for political grandstanding and demagoguery. When they thought about atomic energy at all, they were preoccupied with the "secrecy issue." Witch-hunting made for better headlines and better reelection prospects than did aggressive oversight of the military or formulation of control proposals.

The ignorance of the politicians about what constituted a military secret and what constituted general scientific knowledge and their indifference about observing that distinction led to some ludicrous actions. Senator Millard Tydings criticized the publication of a picture of a cyclotron, which, Lilienthal noted, "had no more to do with atomic weapons than the picture of an electron microscope." Congressman James Van Zandt demanded, and got, a security fence erected around the cyclotron at Berkeley. Lilienthal concluded that the Joint Committee on Atomic Energy actually hindered his goal of increasing public awareness and support for atomic energy because the rest of the members of Congress paid little attention to nuclear matters, since they thought the committee was handling the job. But it was not. The members did not read the commission's reports and probably could have cared less about them. They came to meetings only when there was some political grandstanding to be carried on. The machinery established by the McMahon Act was simply not working.[51]

One of the most fundamental problems faced by Lilienthal and the commission was the inability to get across the idea that there was no atomic "secret" that was in danger of being whisked away. For example, on the eve of the transfer of control over atomic energy from the army to the AEC, a newspaper photographer called the AEC's public relations department and asked whether it would be possible to get a photograph of General Leslie Groves "handing the secret to Chairman Lilienthal."[52] Closely related to the problem of congressional and public ignorance was the problem of "security," which Lilienthal called the "worst part of the job." He found himself increasingly infuriated by the interference of the Joint Committee, which seized on every opportunity to accuse the commission of being lax with regard to security risks. More opportunities for mischief arose when the commission began to distribute scholarships in conjunction with the NRC. It turned out that one of those recommended for a fellowship had been a Communist Party member. Senator Bourke Hickenlooper then took the security hysteria to the extreme of calling for the exclusion from AEC work of all those with "potentially subversive or otherwise objectionable views."[53]

Lilienthal increasingly came to hate himself for the degrading compromises he had to make in order to function politically. He compromised on the AEC fellowship flap and agreed that loyalty oaths rather than full FBI investigations be required of all recipients. Despite these compromises, in May 1949 Hickenlooper launched an investigation into what he called "incredible mismanagement" at the AEC. Among the most damaging things he unearthed concerned the way garbage cans were handled at AEC facilities. The inquiry was, Lilienthal noted in his journal, a "dull, miserable, and humiliating experience." A report issued late in the year exonerated Lilienthal of Hickenlooper's "charges."[54]

Lilienthal thought he had found a solution to the security problems when in 1948 he appointed former Supreme Court justice Owen Roberts to head a security review panel, which also included Karl Compton, Joseph Grew, George Humphrey, and H. W. Prentis Jr. Roberts was left to devise his own procedures, however, and he promptly excluded any right of appeal from his panel's recommendations. Private industry did not offer such a right to a rejected applicant, he reasoned, so why should the government? But recognizing the danger of destroying someone's employability through a negative loyalty evaluation, Roberts recommended that application forms and security questionnaires be consolidated to include a statement in bold print advising that loyalty as well as competence would be considered in evaluating the application. Such a pedantic, limited approach to the problem was not the solution Lilienthal had hoped for when he selected a former justice of the Supreme Court. The commission's security system remained a perverse inversion of American judicial principles.[55]

Lilienthal could take little solace in the company of his scientific colleagues. Few of them shared his daunting visions of the "peaceful atom," and in fact the scientists of the General Advisory Committee and some of his former colleagues in the international control effort rudely dashed those hopes. Early on in his tenure, he was advised by the GAC, "without debate" Oppenheimer recalled, and "I suppose with some melancholy," that "the principal job of the Commission was to provide atomic weapons and good atomic weapons and many atomic weapons."[56] Lilienthal was scolded for his quasi-utopian rhetoric. "Bush bawled me out for the Crawfordsville speech," he reported in his journal. In the speech, he had equated atomic energy with solar energy; the atom, he argued, was a miniature version of the sun and thus was nothing but "a huge atomic-energy factory." Carroll Wilson, Bush's protégé and the general manager of the AEC under Lilienthal, commented that "the power thing was pie in the sky, really. . . . All of the other priorities were higher than nuclear power." Lilienthal recorded in his journal on July 29, 1947, that he "had quite a blow today," after being given a draft of a statement by the GAC that "not only discouraged hope of atomic power in any substantial way for decades, but put it in such a way as to question whether it would ever be of consequence."[57]

Oppenheimer, for whom Lilienthal had the highest respect, dealt the chairman another crushing blow when he severely criticized the very organization and leadership of the commission in the spring of 1948. His complaints were shared by the other members of the GAC and were expressed in what Oppenheimer described as a "rather difficult meeting." He restated the criticisms in a letter to Lilienthal in which he also tried to soothe his friend's hurt feelings. His words revealed the extent of the disillusionment among the scientists about the drift of the government's atomic energy enterprise.[58]

Oppenheimer reminded Lilienthal that when he was wrestling with the question of whether to accept the chairmanship of the AEC, he had told Oppenheimer that he considered international control and secrecy the two most important issues facing the new commission. Those still remained the outstanding problems, Oppenheimer noted, and yet the commission had failed to take any constructive steps in either direction. "We have all looked," he wrote, "to an administration in atomic energy which would combine responsibility with candor, which would be quickly and sympathetically responsive to the needs, the discoveries, and the views of those actively engaged in the work." He had hoped the commission would be "a prototype of an enterprise conducted by the Government in which the free imagination and creative skill of men would find incentive and congenial climate. We looked for a scrupulous willingness in the administration to discover and correct its errors and to admit them publicly; we looked for leadership in policy, in questions of secrecy and security. . . . It was not wrong to look for these things, nor, in my opinion, is it wrong for us to continue to hope for them."[59]

Oppenheimer was pointing out an important weakness of the kind of state science practiced by Lilienthal as well as by Bush and Conant: merely establishing an organization on paper did not mean it would automatically achieve the goals its supporters or even its leaders envisioned for it. There had to be leadership; there had to be willingness and an ability to articulate one's goals clearly, to distinguish them from the competing interests of others who sought to shape the organization in a different way. And, if necessary, one had to fight publicly as well as bureaucratically to achieve those goals. Lilienthal, spinning out fantastic images in his speeches about the bountiful peaceful atom while compromising on the secrecy issue, failed to engage in the hard intellectual and political work that Oppenheimer was calling for. These criticisms not only served to dampen Lilienthal's enthusiasm for his job, they threatened the very survival of the commission. What good was a civilian commission if the only real application of atomic energy was for military purposes? That being the case, and with even the scientists unhappy about the commission's performance, should not the whole enterprise be brought under the control of the military?

Leslie Groves certainly thought so. He continually sought to undermine the credibility of the AEC. After the commission was officially established, he eagerly predicted its failure to the Armed Forces Special Weapons Project and anticipated that the military would have to take over the project. Some members of Congress remained sympathetic to the idea of military control. It played nicely into the demagoguery about security risks: military control presumably meant tighter security and thus a safer republic and enhanced chances for reelection. In April 1948, in fact, Senator Kenneth Wherry introduced a bill to return the AEC to military control.[60]

At least, Lilienthal consoled himself, the commissioners were getting along as a team. Amid all the frustrations caused by congressional demagoguery and military backbiting, Lilienthal took solace in the collegial and cooperative atmosphere that initially existed among the five commissioners. Lewis Strauss was a conservative Republican with experience in investment banking. Sumner Pike was also an investment banker. William Waymack was a journalist and a Republican. Robert Bacher was the lone scientist on the commission, and Lilienthal was the politically independent New Dealer. The ability of men of such differing experience and temperaments to work together sustained Lilienthal's hopes about the usefulness of organized professional expertise in the high councils of the state.

But even this collegiality began to deteriorate as a result of the political maneuverings of Strauss. When the commission voted four to one in favor of providing radioactive isotopes for cancer research to foreign countries, Strauss wanted to explain his dissenting position to Robert Lovett in the hopes of getting the State Department to ignore the AEC's recommendation. "I am far from happy about it," Lilienthal noted. Lilienthal knew little about the courtly, self-made millionaire who had been dabbling and investing on the fringes of American science for several years. At the outset of their tenure together he could little suspect, but he would soon come to roughly learn, that Strauss was determined to make his mark as a powerful and important figure in state science.[61]

Strauss, like his mentor Herbert Hoover, had been born into middling circumstances in life. Both men were forced to struggle in their early lives to attain an education and a leg up. Strauss had been prepared to enter the University of Virginia when he contracted typhoid fever and had to miss the school year. The following year, he was set to reenter, but his father's shoe-selling business floundered in the wake of the 1913 panic. Young Strauss had to pick up the slack and work as a traveling salesman through the rural Tidewater area. In three years he saved the remarkable sum for 1918 of $20,000 and was once again about to begin college at Charlottesville when his mother encouraged him to seek a position working for Herbert Hoover, the popular millionaire-engineer who had turned public servant in the cause of aiding war refugees and feeding starving people in Europe. Hoover was the type of role model Strauss's mother had repeatedly beseeched her son to emulate. Strauss gained an introduction to Hoover by simply requesting an interview at the offices of the new U.S. food administrator. Strauss offered his services without compensation, and on those terms Hoover gladly accepted. The two men began a professional and personal association that lasted until Hoover's death forty-seven years later. These circumstances were made even more remarkable by dint of Hoover's anti-Semitism: "his one fault," an adoring Strauss would later say.[62]

Strauss became Hoover's personal secretary, thereby gaining introductions to powerfully connected men. Doors began to open to this bright, effective, hardworking, and devout young man, who had yet to go to college and who never would. Felix Warburg, who was active in refugee work in addition to his duties as a Kuhn, Loeb partner, offered Strauss a job at the firm in April 1919. Strauss accepted and by 1929, at the age of thirty-three, he had risen up the ranks to a full partnership and a degree of wealth that removed all financial insecurity for the remainder of his life.[63]

During World War II he was in the Naval Reserve and rose to the rank of admiral, a designation that remained his preferred form of address throughout the rest of his life. His work in the office of naval procurement gave him insight into the ways of Washington while also allowing him to forge yet more important connections: with James Forrestal, later secretary of defense, and Sidney Souers, who would subsequently become the executive secretary of Truman's National Security Council (NSC). Souers would also be the conduit through which Strauss would bring his concerns about the hydrogen bomb to the president.

In the fall of 1947 Lilienthal was still laboring under the illusion that Strauss was a reasonable and flexible man who could be dealt with. He met with Strauss to try to prevent the opening of a serious rift over the isotope incident. Strauss was deeply apologetic and offered to resign. Lilienthal dismissed that idea and simply explained how important it was that the commissioners not stab each other in the back by scheming with others in the government to try to undo matters that had already been settled. Strauss became somewhat emotional, calling Lilienthal a "saint" and telling him how terrible he felt. "Don't criticize yourself that way," Lilienthal consoled him, "you just didn't realize what you were doing." Strauss looked at Lilienthal and grinned, saying, "No, I'm old enough; I knew exactly what I was doing." It was a revealing and ominous admission.[64]

In July 1948 Strauss once again found himself the lone dissenter. This time the issue was technical cooperation with the British, which he very much opposed. Britain had lurched so far to the left in his estimation that giving them the "secret" would be tantamount to giving it to the Russians. He asked Lilienthal whether their disagreement should be taken to the president. Lilienthal, noting Strauss's trembling hands during their conversation, rejected the idea, telling Strauss that he seemed "emotionally involved in some way."[65]

Lilienthal soon learned that Strauss had once again gone behind his back to undermine a settled decision of the commission. This time, the admiral had taken his complaints to Senator Hickenlooper, who in turn pressured Forrestal and Bush for an explanation of what was going on. Strauss was now taking his back-channel complaints "all over the place," Lilienthal wrote, and by 1949 the trust and camaraderie that once existed on the commission had nearly disintegrated. "Lewis," he noted dejectedly on the eve of the discovery of the Soviet atomic test, "has made it almost impossible to enjoy the Commission as a family, as we did when we started out, something I worked hard to develop." The discovery of the Russian A-bomb, "Joe I," as the Americans called it, would do nothing to bind up that "family's" wounds.[66]

All of the problems that Lilienthal had encountered during his tenure—congressional demagoguery, pressure from the military, and discord among the commissioners—were soon to be exacerbated to an extreme pitch by the discovery of the Soviet bomb. Lilienthal learned of it in a somewhat disconcerting manner. While vacationing on Martha's Vineyard to recuperate from the grueling weeks he had endured under Hickenlooper's reckless attack on the "mismanagement" of the commission, he was driving with his wife back to their house one evening when along the road he was met by General James McCormack, head of the AEC's Division of Military Applications. Over a kerosene lamp in a darkened room, lending what Lilienthal called a "Charles Addams cartoon flavor" to their discussion, McCormack told him of the readings picked up by the air force's long-range sensors. The American atomic monopoly was over.[67]

Strauss almost immediately began to apply pressure toward the creation of a thermonuclear weapon. His October 5, 1949, memorandum to his fellow AEC commissioners initiated the debate over whether to create a hydrogen bomb. "It seems to me," Strauss wrote, "that the time has come for a quantum jump in our planning (to borrow a metaphor from our scientist friends)—that is to say, we should now make an intensive effort to get ahead with the Super."[68]

Lilienthal had no doubts about his position on the Super. He was definitely against it. It did not even have the potential peaceful uses he imagined the fission bomb to possess. "It is straight gadget making," he said. What he dreaded above all else was the Joint Committee sinking its teeth into the issue and indulging its demagoguery over a new "secret," a new "absolute weapon" that the United States desperately needed to develop before its foes. Once that "plot" took hold, there would be no hope for a rational, intelligent debate. The committee would never understand Lilienthal's arguments that the Super only encouraged an unhealthy and unrealistic reliance on nuclear weapons in American military strategy. The A-bomb had given a "false sense of security," and the Super would only intensify the delusion.[69]

The AEC took up the question and placed it before its General Advisory Committee, which, under the terms of the Atomic Energy Act, had responsibility for advising on technical matters. Oppenheimer had been selected to head this nine-member board in 1947, and he chaired the panel's deliberations on the question of whether to follow Strauss's advice. After three days of meetings in late October 1949, the GAC came out strongly against the idea of developing an H-bomb. The reasons, expressed in a report signed by Oppenheimer, and two annexes, a majority and minority statement, contained the famous and powerful language that denounced the weapon as a tool for "genocide" and "an evil thing" in any light.[70]

A brief but important exchange occurred between Conant and Strauss, who sat in on part of the GAC's deliberations. When Strauss tried to minimize the significance of the committee's deliberations, saying the final decision would be made "in Washington," Conant replied that "whether it will stick depends on how the country views the moral issue." Here Conant's thoughts reached back to the ideas of the Franck Report and represented a notable departure from his own rush to use the atomic bomb in 1945: scientific leaders had to act within some permissible boundary of public legitimacy. People such as Conant and Lilienthal and the GAC members were right to see this issue as a momentous break not just for American science but for American political culture: if the ability to make a weapon was sufficient justification for making it, then scientists would have no political control over their creations. Instead, military appetites would be the only limitation on the size of the American arsenal. The predominance of military influence over such a fundamental political issue in time of peace represented a crippling blow to democratic decision making.[71]

The AEC as a whole voted by a four-to-one margin against development of the H-bomb, later changed to three to two after newly appointed commissioner Gordon Dean switched his vote. This action set off a furious debate within the scientific and political community. Strauss, voting on the losing side of the issue and "more deeply disturbed" than he could describe, found himself "at odds with four men whose patriotism was unquestionable and a General Advisory Committee for whose scientific competence I had a respect verging on awe." He could not understand how a conclusion other than the one he perceived could have been reached. He expressed this view in a November 25, 1949, letter to Truman. "I believe it unwise," he wrote, "to renounce, unilaterally, any weapon which an enemy can be reasonably expected to possess."[72]

This was, of course, the argument advanced in 1940 and 1941 by Lawrence, the Comptons, and ultimately Bush and Conant. But the last two, along with Oppenheimer and others, were convinced that the old arguments no longer applied. Conant, in a less sophisticated rendition of Marx's famously ironic maxim about history occurring twice, first as tragedy and again as farce, noted how the Super debate in 1949 was like "seeing the same film, and a punk one, for the second time." The Super, he believed, "would just louse up the world even more."[73]

Lilienthal was frustrated over the powerful momentum pushing toward a decision favoring development. When he discussed the matter with Truman he warned the president that should the existence of the internal debate become publicly known, there would be terrible pressure from the demagogues in Congress. Truman assured him he would not be bullied into making a politically expedient decision. "I don't blitz easily," he said.[74] But when Truman considered the issue, he adopted the views of the Cold War fundamentalists. The Joint Chiefs of Staff (JCS) produced two written statements on the matter in late November 1949 and in mid-January 1950. In one of them, Omar Bradley argued that it would be "intolerable" for the Russians to get an H-bomb first. The JCS doubted that American restraint would have any effect on the Soviets, and they urged that at the very least research on the weapon should go forward. Bradley rejected the notion that the Super was any more "immoral" than any other weapon. He also relied on what he considered the inevitability of scientific development as an argument in favor of creating the weapon. Truman told his secretary of defense, Louis Johnson, that he felt the JCS arguments "made a lot of sense." The president also heard similar advice from a special three-member committee he had impaneled, consisting of Secretary of State Dean Acheson, Johnson, and Lilienthal, with Lilienthal dissenting from his copanelists' position. Johnson, who believed that a unanimous military judgment of the Joint Chiefs was something the president had to follow, came out strongly in favor of the new weapon. Acheson also favored development, not out of any great enthusiasm for the weapon but simply because, as he told Lilienthal, there was no viable political alternative for the president.[75]

Lilienthal tried to argue that more time was needed to consider the matter. There was no real national emergency that required an urgent decision. No responsible member of the government believed that war with Russia was imminent. Truman himself, in April 1949, in the wake of the Berlin crisis, had stunned Lilienthal by speaking with confident optimism about his expectation of a "general settlement" with the Russians within the next two years, a settlement that would include the outlawing of atomic weapons. Lilienthal further argued that the issue "presented a clear case where the underlying assumptions, policies and plans of the Military Establishment to provide for our defense needed to be examined independently if there was to be substance to the principle of civilian control of atomic weapons by the Commission." If a military judgment was "regarded as the whole answer to the ultimate question, then this definitely removes any notion of civilian participation in a fundamental policy question."[76]

Acheson agreed with Lilienthal's arguments but said that from a political point of view there was no alternative. The pressure from the Congress was just too great. Johnson agreed, telling Lilienthal, "We must protect the president." The trio decided to express their differing views to Truman directly. They proceeded to the White House, where Johnson already had an appointment, and they engaged in a seven-minute discussion of the matter. Lilienthal realized that Truman had "clearly set on what he was going to do before we set foot inside the door."[77] Afterward, Lilienthal met with the GAC members and glumly recounted Truman's decision to a meeting that was "like a funeral party—especially when I said we were all gagged." Someone broached the possibility of resignation; Lilienthal counseled against it. Conant also declined that option, saying later that he "did not want to do anything that seemed to indicate we were not good soldiers and did not do what we could to carry out orders of the President!" Once again Lilienthal chose not to make a "ruckus."[78]

The entire process, marred by shabby political pressures, conducted in a simple, unreflective, and politically unimaginative manner, exasperated Lilienthal. "We keep saying, 'We have no other course,'" he complained to his journal, while "what we should say is 'We are not bright enough to see any other course.'"[79] Lilienthal also could not see any way of publicly contesting the starkly different conceptions of the public interest that had emerged in this debate. Cooperation itself was obviously not the answer to all problems because sometimes the responsible authorities could not resolve their differences. Then what? Lilienthal and his brand of liberalism had no answer. He had never expected the problem in the first place.

By the end of his tenure, Lilienthal was completely disillusioned with the commission. When he told Truman of his intention to resign, he confessed that he had lost all hope in the usefulness of the AEC. "I've tried to make it work," he told the president, "but it really should be re-examined." Its functions seemed to be limited to formulating general policy and public relations. He eventually recommended to the president that a part-time commission and one full-time commissioner replace the current setup. To his close associates he was blunter and bitter. The AEC, he complained, had become merely a contractor for the Department of Defense.[80]

Yet he still clung to his Bellamyesque and Lippmannesque faith in the system evolving its own salvation. Consider this remarkable passage from Lilienthal's journal: "Isn't the whole notion of 'reform' as a fight by one class or economic group against another pretty well a museum piece, without much present relevance? Much of our contemporary 'issues' still use the language and the strategy, but isn't that just a hangover, a legacy of a period when that meant something?" He still believed that "as 'private' business gets bigger, as more things depend upon it, it loses its power, in the sense of arbitrary power. This is not because there is a 'law' that imposes such added responsibility. It is because it is recognized that this vast power over the lives of others carries with it, in the very nature of things as they are in the American ethical climate, a kind of responsibility. Is this paternalism? . . . I don't know that it takes words, tags, to explain it."[81]

Moderate scientific managers and statesmen like Bush, Conant, and Lilienthal had used certain progressive, liberal ideas that seemed to transcend politics. But they had also used those ideas to create political forms and institutions like the cost-plus contract with universities and industry and the government laboratories, and these creations had profound political effects. Just as the ideology of corporate science tried to blur the private institution of the science-based monopoly into a kind of public institution, so too did Bush's wartime management blur the line between public and private organizations.

The collaborative, meritocratic, and nonpartisan models that Bush, Conant, and Lilienthal championed allowed scientists and professionals to exert a greater role in shaping government policy, but they also created new venues for political warfare. Who would control these new institutions of state science? How would they be used? Scientists serving on bodies like the Research and Development Board, the AEC, and the GAC were subject to attack from rival political and administrative factions. These new organizations were hardly the distinguished bastions of elite authority that Jewett talked about. They were in the thick of the internal political conflicts of the state, and those conflicts were not conducted with the collegial, rational purity that people like Lilienthal, and Walter Lippmann and Herbert Hoover before him, had imagined when they conceived of a well-managed, scientific democracy.[82] The opponents of the moderates were hardly responsible, and they were hardly at a loss to come up with words for the goals they wanted to achieve—military supremacy over the Soviet Union—and pejorative tags for the liberal and moderate scientists and administrators who stood in their way.

Bush's and Conant's ideas about a scientifically driven prosperity and a classless meritocracy were clear echoes of Herbert Hoover. But Hoover's notions of progressive, American individualism had one essential condition: peace. His whole conception of the distinctiveness of American democracy was drawn from the horrified contrast he made between war-ravaged Europe and an unscathed America between 1917 and 1920. War and militarism, Hoover knew, wrecked any society through which they passed. Many other Americans knew that as well before 1940. But the 1940s saw a strange transformation of American culture. War was sentimentalized and mythologized; military experience became a noble and even a defining experience of civic identity.

The writings of Bush and Conant were part of that transformation. They helped to normalize militarism by tying the old Hooverian goals of harmonious prosperity and classless meritocracy to an expansive, militarized state. Lilienthal got the first bitter taste of the consequences of that ideological and institutional union. Worse was soon to come.

It was not just the public writings of these three moderates that would influence the historical events of the 1950s. Their silence in the face of the growing militarism in American science and American politics would be just as important. They tacitly ratified the growing power of militant anti-communism in their profession and in the larger political culture. They further ratified a form of professional and political behavior that had been developing since the creation of corporate science in the early twentieth century: experts who worked with the powerful did not openly challenge the policies of their allies and benefactors. They were simply good soldiers.

## CHAPTER 5

## THE BATTLES OVER SCIENTIFIC
## MILITARISM IN THE COLD WAR STATE

Militarization entered American science and American political culture as part of an ideological train that included efforts at recasting the nation's institutional and social structure. Bush, Conant, and Lilienthal believed that even though the ideas and the institutions of state science might be based on military interests, their work would redound to the benefit of the society, making it more progressive, prosperous, and even more democratic. But from out of the Trojan gift horse of state science would emerge a political force I will call scientific militarism that would overwhelm and push aside the political and cultural goals of comparative moderates like Bush, Conant, and Lilienthal.

One of the forces that would overwhelm the moderates was the political and institutional momentum that was building up in the system of national laboratories and state-sponsored science that the wartime leaders had set in motion at the OSRD, the AEC, and the service laboratories. These institutions have received the attention of many historians in recent years, and our understanding of the internal history of state science is richer as a result.[1] These works demonstrate the increasing military domination of American science in an intellectual and not just an administrative sense. Much of this scholarship argues that the fundamental ideological assumption of state science was false: there was not an equal partnership between scientists engaged in independent, basic research, on the one hand, and the military who sponsored and benefited from their unfettered discoveries, on the other. Rather, military imperatives ineluctably channeled the scientists' choice of subjects and methods toward military applications.

An unquestionably coercive effect lurked within these arrangements. Just as in the prewar institutions of corporate science, scientists got to do their basic research with plenty of strings and maybe even plenty of delusions attached. At the Office of Naval Research, Alan Waterman and Captain R. D. Conrad noted that researchers were "entirely free to publish the results" of their work, "but . . . we expect that scientists who are engaged on projects under Naval sponsorship are as alert and as conscientious as we are to recognize the implications of their achievement, and that they are fully competent to guard the national interest."[2]

The question of military control over scientific work is an important one, but it is part of a broader problem: how was it that military interests were allowed to achieve such primacy not only in a profession but also in a political culture in which militarism was feared and there were persistent expressions of concern over the danger of a "garrison state"?[3] We must understand how the ideology of scientific militarism was constructed and how it was used. By examining the ideas of the key authors of this ideology, we can see how their views drew upon the ideas of the corporate scientists as well as those of the wartime leaders. Scientific militarism was an ideology anchored in the idea that scientific innovation was crucial to American progress, prosperity, and, ultimately, national security.

I would like to stress the centrality of certain key individuals—specifically Edward Teller, Lewis Strauss, and, to a lesser extent, Ernest Lawrence— in creating scientific militarism. A powerful institutional architecture had been developing since 1945, which made it possible for ambitious scientists and military and congressional officials to exploit each other for their own gain. But I want to avoid a tendency in some of the scholarship on the history of American science that suggests the sheer existence of certain institutional capacities led to the militarization of American science.

Institutional potentials can lie dormant unless people actively and ener-getically use them for certain purposes. That is what Teller and his allies did. They used particular ideas to carry out their institutional agenda, and those ideas were strikingly similar to the ones that had been used by the corporate scientists and the wartime leaders. The post-1945 militarization of Ameri-can science was not so much a radical departure from prewar arrangements as it was an example of old and familiar concepts being given different emphases and deployed for new purposes.

Posing the issue this way allows me to make a point about the way intellectual and institutional history in this period worked together. The ideas used by scientists to explain their actions were not window dressing used to garnish the raw ambitions that were really driving this system. The old Marxist formula about the institutional and economic "base" creating the "superstructure" of ideas can be abandoned in this instance. It gets things exactly backward. Ideas about the primacy of basic research, the importance of elite collaboration, the realities of technical evolution, and the emergence of an immediate and permanent military threat were not some opiate of the researchers, used to justify the new connections between science and the state. They constituted the intellectual and cultural context within which state science developed and changed. The ideas shaped the actions, and the actions changed the ideas.

To ignore the role of ideas—and key individuals—in the development of scientific militarism is not only to engage in deterministic history (in which the mere *existence* of certain institutions causes change), but it is also to resort to a slightly more sophisticated version of the myth of scientists as dupes who were lured into corruption by the military.[4] While recent scholarship on the history of American science shows how eagerly scientists looked to the state and its money to fund their research, scientists are too often depicted in this literature as the dupes of wily political and military officials who created a "reserve labor pool" of scientific talent that could be used in the event of future military crises.[5]

But scientific militarism was not something imposed on scientists. It was the culmination of a long process of ideological and institutional change that was driven as much by the professional and political ambitions of scientists as it was by the ambitions of military and political officials. The evidence from the postwar period shows how strongly the ideas of the corporate scientists of the 1920s and 1930s and those of Bush and Conant in the 1940s were shaping the new institutional arrangements of the 1940s and 1950s. The administrators of Cold War state science were using many familiar ideas to organize and legitimate their work, such as the importance of basic research as an instrument for achieving prosperity and national security. They also stressed the importance of elite cooperation, and they continued to invoke the idea of evolution as a justification for the changes they were carrying out.

These influences are evident in one institution after another. For example, when the National Science Foundation (NSF) finally got established, Alan Waterman had to fight against congressional budget restraints by emphasizing the military importance of basic science. When the House recommended cutting the NSF's appropriation from a requested $14 million to $300,000, Waterman countered by arguing that "in field after field—aircraft design, jet engine metallurgy, guided missile development, liquid fuel production, military medicine, atomic power—technical progress is seriously delayed by lack of basic knowledge." William Golden advised Truman in a 1950 report that basic research would be "vital to broaden the foundation of knowledge for our military and industrial strength and the public welfare over the longer term." Byron Miller, an official in the Truman administration, urged passage of the NSF legislation, saying it could provide "in a very real sense the first line of national defense."[6]

Ideas about collaboration among basic scientific researchers, the military, and industry were emphasized especially in the AEC, which continued the OSRD policy of contracting its weapons work to universities and industry. "We must maintain," wrote AEC research director Kenneth Pitzer to Allen G. Shenstone of Princeton, "a continuing supply of fundamental knowledge on which technological progress so completely depends, and assure a continuing flow of competent young scientists into industry, university and government laboratories, and the armed forces."[7]

Thomas H. Johnson, the AEC research director in the mid-1950s, had cultivated basic research since the end of World War II. As a member of the Army Service Forces Ordnance Department and as the chief physicist of the Ballistic Research Laboratories, Johnson argued that basic research programs were necessary "to maintain sufficient contact with the civilian agencies which will be making the important progress in physics during the next few years." The "immediate aim" of basic research programs was to "attract first rate nuclear physicists."[8]

These institutional practices further strengthened ideological tenets about the progressive effects of such collaborative enterprises. In a 1955 address before the Northwestern Institute of Technology, Johnson said that "in a rapidly developing civilization such as we are now experiencing one expects an evolution in our educational institutions as in most everything else." Therefore, there should be greater cooperation between government and universities in efforts such as the large accelerator projects. Would this cooperation result in an infringement of academic freedom or an unwarranted expansion of government power? Note the passive voice as well as the Bellamyesque evolutionary imagery Johnson used in answering these questions: "In retrospect these projects can be seen to have arisen to meet the nationally felt need for a higher level of basic research."[9]

One can find numerous examples of this continuing evolutionary enthusiasm among the leading administrators of state science during the 1950s and 1960s. Lloyd Berkner, who served under Bush at the Research and Development Board and later became the head of the Associated Universities, Inc. (the organization that supervised the AEC's Brookhaven research facility), said in 1951, "As the barber-surgeon of the Middle Ages has given place to the medical man of today, with his elaborate scientific training, so the essentially amateur politician and administrator of today will have been replaced by a new type of professional man, with specialized training. Life will go on against a background of social science. Society will have begun to develop a brain."[10]

The army's chief of research and development, A. G. Trudeau, wrote in 1962 that "basic research . . . is the key to technological progress. . . . If yesterday's miracles are today's relics, what an age tomorrow will be, with science as the guide." F. Joachim Weyl, in 1966, celebrated the achievements of the Office of Naval Research as "a striking development in an evolutionary process. . . . Direct government aid through the Office of Naval Research simply multiplies manyfold the available funds to carry American science to new heights in the traditional manner."[11]

The importance of this alliance rippled beyond the universities and the state and into the business world as well. AEC commissioner Thomas E. Murray wrote to Henry Ford II, advising him that the AEC needed to "get many more high caliber men from industry to do a stint in our program and later act as missionaries to industry." If not, Murray warned, "tighter public control would result."[12] Collaboration was the issue, and the major difference between this vision of collaboration and the one envisioned by the corporate scientists was that the state was now assumed to be a legitimate part of this associational arrangement.[13]

Before World War II, Bush had to fight to get his scientific colleagues to accept the notion of a strong alliance between scientists and the military. By the late 1940s and early 1950s, that alliance was taken for granted. The militarized science of the war years, conducted within the ideological framework of corporate science and melded to the institutional structure of state science, now led to the creation of a new ideology: scientific militarism. The proponents of this new mentality insisted that militarism or national security should be the focal point of American political culture and that this was necessary because of the existence of an evolving scientific and technological threat. The answer to that threat, they believed, was the aggressive development of American scientific and technical talent in support of military weapons production.

Edward Teller was instrumental in distilling these ideas about technical evolution and the need for military-scientific collaboration. Teller, in conjunction with Lewis Strauss and with support from Ernest Lawrence, emerged as the leading author and advocate of scientific militarism. It seems inconceivable that the myriad connections between scientists and the state that developed during the 1950s could have been sustained without the radical shift toward permanent war mobilization and the cultivation of a mood of evolutionary, technological crisis that Teller proved instrumental in fomenting. By his zealous advocacy of a thermonuclear weapon he was instrumental in reorienting American policy around massive nuclear weapons, and he helped create the ideological context in which scientific militarism could flourish. He was not the only person who brought this about, but his actions were crucial. What values and ideas guided those actions?[14]

Teller's case reveals a remarkable adaptation of the corporate scientific creed to the military ambitions of the state. He spoke about the importance of innovation and evolution in weapons technology in the same manner that the corporate scientists had celebrated the innovative civilian technologies of the 1920s and 1930s. Like the corporate scientists, Teller and Lawrence both had professional and institutional ambitions that shaped their thought. An "innovative" arms race would give them the opportunity to practice the kind of science they wanted, in the way they wanted.

Teller, of course, had long been an advocate of the Super. His stubborn and irascible devotion to it had interfered with the work he was supposed to have been doing on the fission bomb during the Manhattan Project. Hans Bethe recalled with some bitterness in 1954 that Teller was allowed to work on the Super at Los Alamos because he simply would not do the work that was assigned to him, and the project leaders gave up trying to get him to do the more pressing work on the fission bombs.[15]

Teller, naturally, recalled a nobler pattern of events that led him to become involved in nuclear weapons. In a 1962 memoir, he recorded how he had been jolted from what he considered the complacency of the pure academic scientist in 1932 by a colleague at Göttingen who was abandoning pure mathematical research to return to his native Russia to work on airplane design. "With Hitler on the rise," Teller recalls his friend saying, "we scientists no longer can be frivolous. We cannot play around with ideas and theories. We must go to work." In 1939, when he learned that neutrons could split a uranium nucleus, Teller thought to himself that he "would be unable to continue playing with theories."[16] As the war came to a close, Teller lobbied to have his bailiwick perpetuated. He was disturbed by the example of other scientists who seemed eager to get back to their academic research. He believed scientists were obliged to pursue weapons work at the expense of their merely personal inclinations for "pure" or theoretical work. Teller's postwar writings reveal this sense of obligation, as well as his scorn of those who did not share it.

In 1945, after Hiroshima, Teller was disturbed at the prospect of so many "relieved scientists who wanted no more of weapons work [and who] began fleeing to the sanctuary of university laboratories and classrooms." He tried to push the idea of continued work on the Super and lobbied Norris Bradbury, Oppenheimer's successor at Los Alamos. When Bradbury offered

Teller the position as head of the theoretical division of the lab, Teller agreed to accept only if Bradbury would in turn agree that work proceed on a Super or on the refinement of atomic bombs. Bradbury said neither option was realistic because at the time there was no eagerness in Washington for such efforts. "I took my problem to Oppenheimer, seeking his advice and support," Teller recounts. "I told him about my conversation with Bradbury, and then said: 'This has been your laboratory, and its future depends upon you. I will stay if you tell me that you will use your influence to help me accomplish either of my goals, if you will help me enlist support for work towards a hydrogen bomb or further development of the atomic bomb.'" Oppenheimer quickly replied that he neither could nor would do so. Teller then said he would decline Bradbury's offer and go back to Chicago. Oppenheimer smiled and said, "You are doing the right thing." That same day, Teller and Oppenheimer met again at a party. Oppenheimer asked, "Now that you are going to Chicago, don't you feel better?" Teller did not and said so. Oppenheimer's next comment may have deeply cut Teller's sensitive ego: "We have done a wonderful job here, and it will be many years before anyone can improve our work in any way."[17]

Teller's frustration in these early postwar days became the germ of a resentment that would flourish into the full-blown vendetta he carried out against Oppenheimer nearly a decade later. He would subsequently write that if "men of Oppenheimer's stature had lent their moral support—not their active participation, but only their moral support—to the thermonuclear effort, the United States would have shaved four years from the time it took this country to develop a Super bomb. But the thermonuclear work was given almost no support in the last months of 1945—or in 1946, 1947, or 1948."[18]

In spite of Oppenheimer's opposition, and perhaps because of it, Teller pressed ahead with his advocacy of the Super. In 1949 and 1950 he "urged the feasibility of constructing a hydrogen bomb upon anyone who would listen." Teller claims that he encountered indifference among military people not just about the hydrogen bomb but about fission weapons as well. At an air force conference in Albuquerque in 1949, he and other scientists sought guidance about the types of weapons they should produce. They hastened to remind their clients that a variety of smaller, higher-yield, and more potent bombs could be built. They were not limited to the sizes specified by the capacity of the B-29 bomber. The generals replied, "The bomb we have now is precisely what we need." Their indifference, Teller believed, was the result of "human inertia" that made people accept the familiar and fear the new. It was also the kind of ignorance, or indifference, that had met the scientists who offered their services to the government in 1940 and 1941 and which Vannevar Bush fought against throughout the 1940s.[19]

When the GAC, of which Teller was not a member, advised against creating a hydrogen bomb, Teller considered that decision a blow against himself, as well as against all of the scientists who wanted to work in the thermonuclear field. The GAC's objections to the Super seemed to restrict pathbreakers like him to "minor improvements in the old field of fission" instead of allowing them to follow their inclinations into "the newer field of thermonuclear reactions." Here Teller was echoing an aspect of the culture of the corporate scientists: he was determined that he and his colleagues in the weapons labs be considered not only practical and hardheaded but first-rate scientific talents as well. Scientists would not fully be able to express those talents if they were restricted to being nothing more than glorified technicians.[20]

In his 1954 testimony against Oppenheimer, Teller summarized the general response at Los Alamos to the GAC report against the Super. He interpreted the report as saying that "as long as you people go ahead and make minor improvements and work very hard and diligently at it, you are doing a fine job, but if you succeed in making a really great piece of progress, then you are doing something that is immoral." The Los Alamos scientists became "indignant" and "mad" and determined to work on the Super in spite of the GAC's recommendations.[21]

Teller's rendition of his single-handed crusade against an indifferent or hostile establishment was something of an exaggeration. In the first place, the military was far from indifferent to the expanding menu of fission weapons being suggested by the scientists. In the second place, Teller was hardly the only person making the case for the new weapons. Oppenheimer himself was aggressively lobbying for tactical atomic weapons. He opposed the Super but not the expansion of a different kind of American nuclear arsenal. A less biased observer of events at that time, Walt Whitman, who was one of Bush's successors at the RDB, thought that Oppenheimer "more than any other man served to educate the military to the potentialities of the atomic weapon for other than strategic bombing purposes." Oppenheimer continually stressed the importance of reducing the size of atomic bombs so that they might be used in smaller aircraft, and that meant something other than the Strategic Air Command's (SAC) massive bombers. But Teller's narrative of events surrounding the Super seemed to require an inert bureaucracy that would be seized and swept away by his enthusiasm for technological innovation.[22]

His narrative also rested on the idea that the arms race represented an evolutionary struggle that the scientists had to wage aggressively in order to stay ahead. "To my mind," Teller once said, "the distinction between a nuclear weapon and a conventional weapon is the distinction between an effective weapon and an outmoded weapon." The danger existed that one's competitors could achieve an evolutionary breakthrough, thus leaving the complacent nation in a vulnerable position with only outmoded weapons. Here again there are echoes of the corporate scientists and their arguments that innovation was necessary to achieve primacy in the patent field so as to block out potential competitors from a particular market. Teller used these arguments, of course, not simply as part of a case for defending a market share but as an appeal for national defense and national survival. If the Soviets developed a Super before the United States, it would put the capitalists "out of business" in the most extreme sense.[23]

By putting his arguments in such extreme, even melodramatic terms, Teller was able to gain more attention and support for his ideas. When the Soviet atomic test occurred in August 1949, he was able to link his ambitions and grievances to a potent strain of Cold War politics. He gained the support of allies like AEC commissioner Lewis Strauss and Senator Brien McMahon, who had turned into a fanatic anticommunist and an advocate of limitless nuclear proliferation.

Another profound similarity between Teller's rhetoric and that of the corporate scientists of the 1930s was that Teller's arguments, if accepted, left no room for alternatives. If there was indeed an evolutionary race afoot, with one's competitors likely to win the prize, and if the consequences of an adversary's victory were as dire as Teller and other Cold Warriors predicted, then the United States had no choice but to enter that race with all the energy it could muster. The corporate scientists of the 1930s had offered a rigid dichotomy between those who would support "progress" and "prosperity" through the corporate system and those benighted few who opposed it. Teller and the Cold Warriors made the stakes of their arguments much higher. One was either for or against national survival. One was either for or against the defense of one's country against a mortal threat from an implacable foe. If one failed to subscribe to this ideological vision, one was not simply wrong or foolish but a traitor.

Teller, remarkably, was able to cloak his aggressive political agenda in the language of scientific nonpartisanship, another key element of the corporate-scientific creed. In a March 1950 essay in the *Bulletin of Atomic Scientists* titled "Back to the Laboratories," Teller wrote, "It is not the scientist's job to determine whether a hydrogen bomb should be constructed, whether it should be used, or how it should be used. This responsibility rests with the American people and with their representatives."[24]

Teller was not alone among scientists, of course, in pushing his hydrogen weapon. For example, Lawrence's protégé Luis Alvarez developed a commitment to the thermonuclear project that was based, like Teller's, on a mixture of political conviction, professional self-interest, and emotion. He shared Teller's and Strauss's view that it was unthinkable that the United States would not have a hydrogen bomb while the Soviets possessed it, and both Alvarez and Lawrence felt inspired to press for development of the Super in the fall of 1949, shortly after Truman announced the Soviet atomic explosion.

On October 6, 1949, Alvarez broached the subject to Lawrence, who did not need much persuasion, and they agreed to bring their case to Washington via Los Alamos and Edward Teller. As the leading authority on the Super, Teller was certainly the person from whom the two Berkeleyites could obtain politically useful information about what a thermonuclear project would entail. When Teller told them of the importance of obtaining tritium, a heavy isotope of hydrogen that could only be made artificially, Lawrence and Alvarez realized that they suddenly had a rationale for a new outpost of state science. The production of tritium was a complex process. A large new facility would be needed, and, as Alvarez and Lawrence told AEC officials when they arrived in Washington, the people at Berkeley "had a demonstrated capacity for rapidly building large scientific systems." The AEC "seemed to be interested," Alvarez recalled in his memoirs, but the chairman, David Lilienthal, was not.[25]

Alvarez could not understand why anyone would be opposed to the Super. Before the GAC meeting, the only opponents to it that he encountered were people such as Paul Fine, an administrative assistant in the Radiation Lab. Fine "was the first person I had met since the Russian bomb went off who was not enthusiastic about the problem of building the Super weapon," Alvarez told the AEC's Gray Board in 1954. "I attributed this to the fact that he had all during the war and was still then sort of an administrative assistant and I put him down as a person with essentially no imagination and discounted this."[26]

David Lilienthal, according to Alvarez, threw a childish fit and "turned his chair around and looked out the window and indicated that he did not want to even discuss the matter. He did not like the idea of thermonuclear weapons, and we could hardly get into conversation with him on the subject."[27] He did no better with Conant. Shortly after the Soviet explosion, Alvarez was riding in a car with Lawrence and Conant, and "Dr. Lawrence

was trying to get a reaction from Dr. Conant on the possibility of radiologi-
cal warfare and Dr. Conant said he wasn't interested. He didn't want to be
bothered with it." Alvarez claimed to "have the strong recollection that
Dr. Conant said something to the effect that he was getting too old and tired
to be an adviser on affairs of this sort. He said, 'I did my job during the
war' and intimated that he was burned out, and he could not get any
enthusiasm for new projects. So when Dr. Conant disapproved of the hydrogen bomb, I interpreted it in light of that conversation."[28] Nerds, petu-
lant men, or burned-out cases were the only people, Alvarez believed, who
could be opposed to a hydrogen weapon.

Alvarez was most astounded by Oppenheimer's objections. In late October 1949, during another trip to Washington, Alvarez had lunch with Oppenheimer, and the former Los Alamos director explained his objections to
the Super. "The main reason he gave," Alvarez wrote, "was that if we built a
hydrogen bomb then the Russians would build a hydrogen bomb, whereas
if we didn't build a hydrogen bomb then the Russians wouldn't build a
hydrogen bomb. I thought this point of view odd and incomprehensible."[29]

Oppenheimer objected to the way Teller, Lawrence, and Alvarez posed
the issue. He did not see the Super in 1949 as the exact equivalent of the
atomic bomb in 1939: something that had to be developed to prevent a
similar attack from a warring adversary. He believed that the wartime attitude had to be abandoned. It need not be American policy, he argued, that
simply because a weapon could be developed, it must therefore be built. It is
easy to understand Alvarez's reaction since Oppenheimer was challenging
the basic evolutionary model that many scientists had been using to think
about nuclear weapons.

But to Oppenheimer the possible creation of a hydrogen bomb was a
political problem and not an inevitable act of technological evolution. In an
October 21, 1949, letter to Conant that would be used against him in 1954 as
an example of his insufficient devotion to American offensive capability,
Oppenheimer wrote caustically of Teller and Lawrence as "two experienced
promoters" whom he resented for undermining the GAC's deliberations by
lobbying Senator McMahon and members of the "military establishment."
Worse still, the efforts of the promoters seemed to be having an effect: "This
thing," Oppenheimer wrote to Conant, "appears to have caught the imagination both of the Congressional and of the military people, as the answer
to the problem posed by the Russian advance. It would be folly to oppose
the exploration of this weapon. We have always known it had to be done;
and it does have to be done. . . . But that we become committed to it

as the way to save the country and the peace appears to me full of dangers."[30] But the Super advocates could not conceive of the kind of political consideration Oppenheimer was proposing. As Robert Gilpin noted, the Super's promoters could see this only as a scientific and technical issue: they thought the Oppenheimer-Conant position was in a sense untrue to the spirit of science because it allowed nontechnical considerations to intrude on the question of whether a particular choice should be made.[31]

The two sides of the Super controversy represented not so much a disagreement over the hydrogen bomb per se as a disagreement over entirely incompatible ways of understanding political conflict and social change. The evolutionary, deterministic strands of thought that had been present in the corporate science ideology since the 1920s were becoming increasingly stronger over the course of the Cold War. Notions of controlling and properly managing scientific change—which had also been part of the culture of corporate science—were being weakened to the point that they were no longer considered legitimate positions subject to debate but were merely aberrant heresies from the new dominant creed. Robert Gilpin wrote in 1962 that "the scientist's integration into political life has been inhibited by the scientist's ambiguous image of himself."[32] We have seen such ambiguity in the thought of people like Bush, Conant, Lilienthal, and even Jewett. But there was very little ambiguity in the minds of Edward Teller, Ernest Lawrence, and Luis Alvarez, the scientists who successfully integrated themselves into political life in the 1950s. They were received with open arms by the political guardians of the republic on Capitol Hill. The ambiguous scientists were being pushed to the margins of this policymaking process. An institutional momentum was gaining strength, but the driving force behind that momentum was decidedly human and decidedly political. To achieve their political and institutional goals, Teller and his associates responded by strengthening their own institutional base through the creation of the Livermore Laboratory and by attacking the other scientific leaders who opposed their policies.

The resentment of the Super enthusiasts began to focus on the leaders of the GAC, Oppenheimer and Conant. In March 1950, Teller wrote to Joint Committee staff member William Borden to complain that "the attitude of the members of the GAC has been a serious difficulty in our recruiting efforts. . . . A man like Conant or Oppenheimer can do a great deal in an informal manner which will hurt or further our efforts." Lawrence also bad-mouthed the commission to members of the Military Liaison Committee, saying that the AEC would not be capable of generating the tritium required for the Super.[33]

Lawrence, the quintessential "nonpolitical man," had a straightforward institutional interest in attacking the GAC's perceived opposition to the Super. He had plans for a large accelerator, the Materials Testing Accelerator, or MTA, that would be used in producing tritium. The MTA, which was ultimately intended for Weldon, Missouri, was designed to be over five hundred yards long. Lawrence had a prototype built at Livermore, California, near the Berkeley campus, and the services of a corporate powerhouse, Standard Oil of California, were used to assist the Radiation Laboratory staff. But the project was never brought to fruition. Technical difficulties, along with delays and the sudden availability of deuterium and tritium from other reactors, killed off the project. The MTA was, in the words of Herbert York, "the first time Lawrence had, in effect, overreached himself in a big way."[34]

Lawrence pressed on in spite of congressional resistance, and he used the familiar language of corporate science. In an April 1951 hearing before the Joint Committee on Atomic Energy, Senator Hickenlooper asked whether Lawrence felt "this is just a gadget to play with" or whether "it is something that may turn up something important." Lawrence cited the "unpredictable value of fundamental science." One could not anticipate the results of research, but the potential gains were too great to risk missing out on them.[35]

But the failure of the MTA project did not mean that Lawrence abandoned efforts to establish a new state-sponsored laboratory under his purview. His ambitions fused nicely with Teller's resentments to form a potent cabal in favor of a second weapons laboratory in addition to, and in competition with, Los Alamos. Teller had continuously lobbied the AEC for a second nuclear weapons laboratory. Even after the decision to build the Super, it was, in Teller's words, "an open secret, among scientists and government officials, that I did not agree with Norris Bradbury's administration of the thermonuclear program at Los Alamos." He was even more incensed when Bradbury did not appoint him director of the project.[36]

Teller left New Mexico and returned to the University of Chicago. But he did not go quietly; he was able to get his complaints into the ears of important people in Washington, namely, Senator Brien McMahon and McMahon's staff aide, William Borden. Even more important, Teller spoke with the chief air force scientist, David Griggs, formerly of the Rand Corporation, and General Roscoe Wilson. He told them not enough was being done to exploit the recently devised Teller-Ulam theory of thermonuclear explosions. These arguments, coming from the author of this innovation, were heeded. But the air force may have had its own agenda for heeding Teller's criticisms. It was, as Herbert York noted in his memoir, calling for "new institutional arrangements" to conduct thermonuclear work and was considering the creation of its own weapons laboratory. This impetus, York surmised, may have added to the pressure on the AEC to create a second lab.[37]

Teller's consistent criticisms and his continuous calls for a second laboratory were more annoying than persuasive to the AEC. The commission had considered and rejected the idea several times. The GAC accepted Bradbury's assertions that a second lab was not necessary and that its creation would only lead to the neglect of Los Alamos. But the notion did not fade away. Thomas Murray, one of the AEC commissioners, had asked the Radiation Lab director for his views on the need for a second lab. Lawrence, in turn, sent York to canvass opinion among those knowledgeable about the thermonuclear effort, and he reported back to his boss that there was a firm consensus in favor of building the lab. Lawrence endorsed the idea, and he had the perfect location in mind: the site of the abandoned MTA pilot project at Livermore, California.[38]

Teller even tried to invoke some good old-fashioned capitalist principles in support of his argument: competition was good. Two labs would better serve their customer than would one. But, perhaps in a similar business spirit, the government agreed to the new lab only when it was able to get it on the cheap. It was Lawrence who provided the convincing proposal: he offered to staff the new lab with people already on his payroll, thus giving the AEC an inexpensive way to open a second lab without also having to raid the talent at Los Alamos.[39]

The Livermore National Laboratory was approved by the AEC in June 1952, and it opened in September. It was imbued with the spirit of Ernest Lawrence, rather than the single-minded hawkishness of Edward Teller. This spirit was embodied in the selection of York as the first director. York's selection, according to his memoir, was entirely Lawrence's choice; Washington seems not to have had much say in the matter. After having York draw up some plans for the structure and operation of the new lab, Lawrence asked his young protégé if he would run the new venture. "After only an overnight hesitation, I told him it was worth a try, and he simply instructed me to do so. It really was that casual; no search committee or any of the other procedures to which we are now accustomed." York "clearly understood and firmly agreed with Lawrence's approach to 'big science,'" and when the young new director showed his plans to Lawrence, "he always warmly endorsed" them.[40]

This new lab provided Lawrence a way to revive the glory days of his wartime work and perhaps even to recover the more innocent energy of the early Radiation Laboratory. York believed that Lawrence considered his work on the Super to have been "not only a matter of duty but . . . a personal opportunity" to regain the feeling that "you were really part of a great movement, doing things which were interesting and consequential."[41] The Livermore Laboratory, like the early Radiation Lab, was a place where young men worked in a relatively unstructured way, in an almost communal undertaking. York was not even given an official title as director. "Lawrence firmly believed," York wrote, "that if a group of bright young men were simply sent off in the right direction with a reasonable level of support, they would end up in the right place." Teller, not surprisingly, was still dissatisfied. The new lab's mandate was not as specific or as aggressive as he would have liked. He initially swore off any involvement with the project, but he was given a role as an all-purpose consultant with an important, although never used, veto over lab projects.[42]

As director of Livermore, York continued Lawrence's practice of engaging in exciting, state-sponsored work while also pursuing fundamental science. His sense of the lab scientists' mission was similar to Teller's conception about the role of Cold War state scientists: they should be engaged in research geared toward the highly practical and patriotic goal of developing new weapons for the military. "Our working philosophy," York wrote, in terms that could as easily have been spoken by an ambitious corporate researcher in the 1920s or 1930s, "called for always pushing at the technological extremes. We did not wait for higher government or military authorities to tell us what they wanted and then only seek to supply it. . . . We were completely confident that the military would find a use for our product after we proved it, and that did indeed usually turn out to be true."[43] York recalled to an interviewer that Lawrence believed that "instead of just picking up what other people are doing, we should do things different—for the sake of doing things different." This philosophy guided many of the lab's projects, in which Livermore was "not just taking what Los Alamos was doing and trying to do it better, but trying to do something substantially different."[44]

This approach struck York as the best way of assuring "American superiority in nuclear weaponry" while also providing "the kind of intellectual stimulus and prospect for adventure that young scientists usually find only in basic research. . . . This approach meant that the laboratory leadership had to engage in a continual effort to sell its ideas, to anticipate military requirements, and to suggest to the U.S. military ways in which its new designs could be used to enhance preparedness and better support our general nuclear strategy." York made it clear that the lab would be initiating new ideas and new technologies. It would not simply be filling production orders from the Pentagon. "If we had waited for Washington to tell us exactly what was needed in terms of dimensions, yield, special output, or other technical parameters, such selling would not have been necessary, but that is not the way we went about our business at Livermore, nor did they do things that way at Los Alamos, especially not after we brought competition onto the nuclear scene."[45]

Once again we can hear echoes of the corporate scientist's credo: basic research in a practical context. But more important, this institutionalization of the ethic of the corporate scientists at Livermore and the other state-sponsored laboratories provided the government's network of scientific institutions with a political significance they had never enjoyed previously. Once the idea of a perpetual, evolutionary arms race had been accepted, as it had been with the decision to go ahead with the Super in 1950, it became crucial to guarantee the perpetual production of new, evolutionary breakthroughs in weapons technology.

The scientists in the laboratories, and the laboratories themselves, were now a crucial and perhaps essential component of the American Cold War arms race. The labs had to be maintained and strengthened so that they could protect the United States against some surprise technological breakthrough by the Soviets. But because their mission was to initiate their own weapons projects, they were not simply allowing the United States to keep up with Soviet innovations. They were themselves driving the arms race to greater and more dangerous heights.

Americans had to evolve their technology because they were caught in a merciless evolutionary process. Thus throughout the 1950s, as growing concerns about fallout and the arms race were expressed by the public, Cold Warriors argued that the continuation of testing and proliferation were necessary in order to continue the search for "cleaner" weapons. Lewis Strauss also argued that if the role of the laboratories was weakened, the United States would be in a poor position to discover the new breakthrough in weapons technology which the Soviets would surely be searching for themselves. It was no exaggeration to say that while the laboratories were necessary in the face of the arms race, the arms race itself was indispensable to preserve the national laboratories.[46] Also, this collaboration between basic scientific researchers and military customers strengthened the notion that such collaboration was necessary and natural. This was how science was done.

Here was a curious triumph of the kind of prestige and status Vannevar Bush had wanted to create for state scientists during World War II. Scientists such as Teller, Lawrence, and Alvarez had certainly formed powerful alliances with the military and the political elite: their interests meshed so nicely with the military, particularly the air force, that the scientists could hardly see how anyone could rationally call for restraint or arms control, as did Bush, Conant, and Oppenheimer. But Bush, in particular, as we saw in his 1949 book *Modern Arms and Free Men*, envisioned scientists as far more than partners in arms production. He wanted scientists to be partners in the formulation of a new governing ideology.

But the ideology he advanced in his book had no place for political conflict between different interests or philosophies. All conflicts and problems were simply "managed" away with good organization and team spirit. The faction of scientists which Bush, Conant, and Oppenheimer faced in the mid-1950s was gripped by an aggressive, militant philosophy of arms proliferation and anticommunism in which ideas of fair play, organizational efficiency, and teamwork were nonexistent.

Throughout the early 1950s, the moderate state scientists continued to try to advance their views from within a system increasingly dominated by the Cold Warriors, and they desperately tried to preserve their position as dependable "insiders." In the early 1950s Bush and Oppenheimer in particular made what amounted to a last stab at bringing their professional and political philosophies to bear on government policy. But the limitations of their arguments, the closed-mindedness of the political and military elite, and the extreme militarization of the state and of American culture made their effort seem quixotic. Like many such efforts, it had its admirable aspects that deserve our attention, especially on Oppenheimer's part: he sought to subject scientific and military policymaking to public argument and conflict. It would be this vision of scientific politics, I will argue, even more than Oppenheimer's specific positions on the Super, that would lead the government to attack and to try to destroy his reputation.


By 1950, Robert Oppenheimer had come a long way from the man who had callously joked with Conant in 1945 about "bombs away by the first of May." The "hero" of Los Alamos, who had rejected the activist role proffered by the Franck Report, had become an extremely disturbed observer of American scientific and military policy. One of the explanations for this change can be found, I believe, in the similar transformations that were taking place in the minds of Bush and Conant, especially Bush, who, unlike Conant, remained active in scientific advisory roles during the early 1950s. These fellow "statesmen" of science also developed profound misgivings about American nuclear weapons policy, and their opinions exerted an important influence on Oppenheimer's own thinking.

Oppenheimer's political battles throughout the early 1950s with the air force, with Lewis Strauss, and with the Berkeley cabal have attracted keen historical interest. But these clashes were indicative of a fundamental ideological schism within the community of state scientists, and Oppenheimer's security record became the focal point of that conflict. That schism, more than the details and personalities involved in the Oppenheimer episode, will occupy my attention in the next chapter. But I do not wish to ignore or minimize the importance of the crucial factor behind the Oppenheimer affair: Lewis Strauss's fanatical hostility toward Oppenheimer. The Cold War extremists might well have seized control of state science even if Strauss loved Robert Oppenheimer and merely agreed to disagree with him. But he did not, and the ways his hatred and suspicion shaped the events of this period cannot be ignored.

Strauss's antipathy, like all irrational hatreds, undoubtedly had some psychological basis that lies beyond historical analysis. His biographer Richard Pfau observed that in every instance when Strauss disagreed with someone on a matter deemed important by Strauss, that disagreement invariably blossomed into bitter enmity. Herbert York, who was an ally of Strauss and Ernest Lawrence in the 1950s, would later run afoul of the admiral when York dismissed a scientific initiative Strauss wanted the government to champion. He later observed in retrospect, "Lewis was absolutely disinterested in bad news and didn't want to hear it."[47]

There were important self-evident differences between Strauss and Oppenheimer that laid the groundwork for the ill will that developed: one man was a devout Jew, the other a Jew who wore his faith lightly and liberally sampled beliefs from the religions of the world; one was a man of conventional mores and strictly conservative political views, the other had been a romantic fellow traveler who married one former Communist and conducted an adulterous affair with another.

Strauss's deep personal antipathy toward Oppenheimer began in 1947, after the latter used Strauss's offer of the directorship of the Institute for Advanced Study, where Strauss was a trustee, as a bargaining chip with other institutions. Oppenheimer took three months to respond to the offer, and Strauss bitterly resented being used.[48] Another, far more spectacular provocation occurred when Oppenheimer ridiculed Strauss in public in 1949 during a Joint Committee hearing over the question of whether radioactive isotopes should be shipped to Norway.[49] None of these petty personal grudges on Strauss's part would likely have inspired him to wreck Oppenheimer's career had not the two also disagreed over the Super and the role of scientific experts in government and also if the organizational mechanisms were not in place to allow Strauss to carry out his attack in secret. But the grudges added the harsh bite of intense hatred to Strauss's deeply felt policy positions.

As a consultant to the Joint Committee after his resignation as an AEC commissioner, Strauss still had access to official documents, and the papers he read concerning Oppenheimer's views greatly disturbed him. In an October 1950 meeting at Los Alamos on the Super, for example, Oppenheimer argued that the weapon seemed to be a technical impossibility. Teller argued strongly that a determined and concentrated effort would inevitably bring success. The theoretical work done thus far, however, indicated that a thermonuclear reaction could be initiated but that it could not be sustained. Oppenheimer had reflected this skepticism in the December 1950 long-range capabilities panel he chaired. The panel's report warned that "only a timely recognition of the long-range character of the thermonuclear program will tend to make available for the basic studies of the fission weapon program the resources of the Los Alamos laboratory."[50]

When he read Oppenheimer's report, Strauss believed the former Los Alamos director was stalling the project. He suspected the most sinister of motives. Was Oppenheimer, in fact, a Soviet agent bent on killing the American Super in its cradle? And what of Oppenheimer's resistance to the notion of creating a second weapons laboratory? Strauss took his fears to Gordon Dean in February 1951. Oppenheimer was "sabotaging the project," he told the new AEC director, reading from a memo. He urged that "something radical" needed to be done: Oppenheimer had to be removed from his position of prestige and influence in state science. At the conclusion of the meeting Strauss portentously threw the memo into the fire. In that same month, he drafted a memorandum to Truman, which does not seem to have been sent, in which he warned that Oppenheimer was like "a commander who did not want to fight." This reluctance, Strauss warned, might have caused the United States to fall behind the Russians in the race for the Super.[51]

Late 1950 seemed to be a low point in the prospects for a Super. But only a few months later, Teller and Stanislaw Ulam would arrive at a new theoretical approach to the weapon, the use of X-ray implosion. Ulam, at first, was skeptical of the new ideas Teller advanced. "Edward is full of enthusiasm about these possibilities," Ulam wrote John Von Neumann; "this is perhaps an indication they will not work." But they did work.[52]

In addition to Strauss, Oppenheimer also attracted the influential hostility of the air force in the early 1950s, when, in connection with his doubts about the Super, he began to advocate the use of "tactical" fission bombs. Greater attention to fission bombs, he believed, might steer the military away from its reliance on the Super as the sole deterrent to Soviet aggression. The sound of the former director of the Los Alamos laboratory criticizing the government's reliance on the Strategic Air Command and its airborne thermonuclear arsenal was music to the ears of the army chieftains who wanted to "bring the battle back to the battlefield." The arguments Oppenheimer marshaled against the thermonuclear school were powerful ones: Would the United States be prepared to demolish nearly all of Europe in the name of repelling a Soviet invasion? Would, for example, a Soviet incursion from East to West Berlin be sufficient to trigger the use of the Super in Europe, which, of course, might lead to a full-scale exchange between Russia and the United States? To Oppenheimer, these questions underscored the excessively dangerous, sledgehammerlike quality of the American deterrent as envisioned by the H-bomb enthusiasts. It permitted no flexibility, and it was simplistic in its assumptions about the willingness of an American president to cause the death of tens of millions of Europeans in the name of defending them.[53]

Oppenheimer believed that tactical nuclear weapons, with their less gargantuan destructive power, might provide a more realistic and thus more effective deterrent against any potential Soviet aggression. But it augured a diminished role for the Strategic Air Command and its thermonuclear arsenal, and that led to some ugly confrontations. In 1952, Oppenheimer called David Griggs, the chief scientist for the air force, a "paranoid," after the latter intimated that he considered Oppenheimer to be sympathetic to the Soviets. Oppenheimer's relations with Thomas Finletter, the secretary of the air force, were just as icy and strained. When a dinner was arranged to include Oppenheimer and Finletter to smooth things out between them, Oppenheimer behaved rudely and contemptuously to the secretary throughout the evening.[54]

The deterioration of Oppenheimer's relations with the military elite coincided with his increasing prominence in advisory or study panels. In the summer of 1951, for example, the army subsidized a study at the California Institute of Technology called Project Vista. It was intended to devise recommendations about the defense of Western Europe, and Oppenheimer was enlisted to draft the report's final chapter on the use of tactical nuclear weapons. He recommended the creation of a Tactical Air Command in Europe, armed with fission rather than fusion bombs. He also advocated that Los Alamos and the other laboratories move away from their emphasis on thermonuclear weapons and instead concentrate equally on strategic nuclear weapons, tactical nuclear weapons, and the Super.[55]

In December 1951, Oppenheimer was even able to make his arguments directly to the head of the North Atlantic Treaty Organization (NATO) command in Europe, Dwight Eisenhower, who reportedly thought the suggestions were sound. But air force officers lobbied the Vista participants during their discussions in the hopes of urging them to reject the army-Oppenheimer agenda. When the project's report was released they even suppressed it and retrieved the copies of the report and the underlying documents that had been in the possession of the participants.[56]

Vannevar Bush matched Oppenheimer stride for stride in his criticisms of military policy. The two served together on a presidentially appointed panel on arms control and disarmament, officially titled the Presidential Panel of Consultants on Disarmament. The panel also included Allen Dulles and two other proponents of international control, John S. Dickey and Joseph E. Johnson. The panel, which was created in April 1952, produced an important report in January 1953 to incoming President Eisenhower, detailing suggestions about the way American weapons policy needed to be changed. But one of the most controversial recommendations was that the November 1952 test of the first American thermonuclear device be postponed. This proposal was actually the brainchild of Bush, who in the early 1950s became increasingly vocal as an insider critic of the arms race.

Bush believed it was wrong to carry out such a test in the waning days of a lame-duck administration. The incoming president would be faced with the reality of a workable thermonuclear weapon, a weapon that would draw intense political and institutional enthusiasm, making it almost politically impossible for the new president to oppose. Postponing the test, Bush believed, was crucial to sustaining any hope of controlling the arms race. For this was not just one more test of a nuclear weapon. It was a profound political act that required public debate. "Let us postpone the test," the panel recommended, "if such a decision can be understood, explained and properly supported."[57] The public would have to be involved, and the political, military, and scientific leaders of the state would have to take responsibility for explaining and justifying the choices surrounding the development of this technology.

Abjuring a test was also, Bush believed, the last best hope of forestalling a runaway arms race. "That test," Bush told Robert Oppenheimer's interrogators in 1954, "ended the possibility of the only type of agreement that I thought was possible with Russia at that time, namely, an agreement to make no more tests." It would have been a self-policing agreement, Bush pointed out, which would require neither trust nor verification, thus making it politically palatable to both sides. Any thermonuclear test would have been detectable with long-range sensors, and if the Soviets violated the ban the United States would immediately know about it, permitting it to resume its own tests. "I think history will show that was a turning point," Bush noted in 1954, "when we entered into the grim world we are entering right now." He added that "those who pushed that thing through to a conclusion without making that attempt have a great deal to answer for."[58]

The final report of the panel, "Armaments and American Policy," was not completed until after January 1953, and when it was presented to the Eisenhower administration the new president had the opportunity to read a powerful indictment of the reckless proliferation and belligerence advocated by the Cold Warriors. It emphasized that the United States was in an arms race in which little heed was given to the purpose or the consequences of a continually expanding American arsenal. American superiority, the report pointed out, would have little significance if the Soviets had simply "enough" weapons of their own. Both sides would be able to inflict devastating nuclear attacks on their adversary but only "at the gravest risk of receiving similar terrible blows in return. . . . A world of this kind may enjoy a strange stability arising from [a] general understanding that it would be suicidal to 'throw the switch.'" But "a world so dangerous may not be very calm."[59]

These criticisms and recommendations revealed the best as well as the worst of moderate state science. They reflected an awareness of the need to balance preparedness with the larger social goal of protecting the nation and the world from the dangers of thermonuclear weapons. They revealed an awareness that choices existed with regard to nuclear weapons: that there were no self-evident truths compelling proliferation, and Americans had to do more than sit back and let the supposedly evolutionary course of science and technology work its magic. Bush was loosening the powerful grip which the evolutionary trope had on his mind. People could, and indeed had to, think, choose, and act with regard to the development of these weapons. But as Bush's career in the 1950s made clear, the leading moderate state scientists crippled their political effectiveness by muting their voices in the public realm and by trying to cling to the ideological structure of the

Cold War while calling for significant changes in the organizational apparatus through which that war was waged. They advanced many strong ideas in their report, but they restricted the expression of those ideas to the narrow and increasingly secretive world of nuclear policymakers.

The 1952 disarmament panel can at least be credited with injecting the ideas of the Franck Report back into the internal discourse of the government, if not into public debate generally. The panelists urged that the government adopt "a policy of candor toward the American people," particularly with regard to the effects of nuclear weapons. Such candor was essential, they believed, if the nation's citizens were to be engaged in this issue. "The more responsible the citizen . . . the more he is likely not to pay full attention to the problem of atomic weapons as long as present security restrictions are enforced." An ordinary citizen would "naturally hesitate to attempt a judgment on any matter on which he knows himself to lack important information; he will tend to leave the problem to those who know the facts."[60]

This was a line of argument that seemed to lead along an entirely different path from the one the leaders of state science had been treading ever since World War II. The moderates began to realize that a healthy society could not preoccupy itself solely with the accumulation of military power and the enforcement of military standards on civil institutions. And military questions themselves could not be completely excluded from consideration and debate by the general public. Such concerns were nowhere to be found in the writings or the actions of the scientific militarists. For Strauss, Lawrence, Teller, and their allies, the American military had become the American state. Its interests seemed to obliterate all others in their minds.

For Strauss, protecting secrets seemed to be the raison d'être of state science. In his memoirs he explained that he saw his role as an AEC commissioner as something akin to a national fiduciary. If the directors of a bank discovered that a teller had formerly been an embezzler, would they not discharge the man? "Our responsibility as Atomic Energy Commissioners was of a far higher order," he wrote. "As bank directors we would be responsible to our stockholders and depositors for nothing more important than money, but as Commissioners, we were fiduciaries of the whole American people for the safety of their lives and the defense of our government."[61]

In Strauss, Bush and Oppenheimer had an opponent who had an extremely narrow view of the entire relationship between government, the military, and scientific experts. Experts were not partners to the governing elite, they were instead merely the servants or fiduciaries of the state; they were to protect the interests of the government (as defined by Strauss), not challenge them. Bush wanted scientists to be important shapers of government policy. Strauss wanted them simply to be instruments of those policies, and Bush did not know how to fight against this redefinition of the professional role he had carved out for himself and other leading scientists. Had not he, after all, continually stressed the collaborative, cooperative nature of scientific expertise in the government? Now that government policy was moving in a direction to which he strongly objected, how could he express his views without being an adversarial critic and losing his position as a valued insider? Instead of taking on Strauss and Teller directly, he aimed at the margins of state science. His recommendations for change were begun reluctantly, they were invariably mild, and they were usually confined to organizational matters. Bush always made it clear that he was a friend of the military, and he thus protected himself against the kind of ruthless counterattack that Oppenheimer received.

While he avoided a public clash over the fundamental issues of Cold War policy, Bush can at least be credited with persistence. He had been propagating his ideas about strategy and military reform in venues such as the *Reader's Digest*, where he wrote in January 1951 that "we need to think of the A-bomb as a tactical weapon to stop great armies." In the inauguration of a series of radio addresses sponsored by the Committee on the Present Danger, Bush insisted that "we cannot count indefinitely upon strategic bombing as the sole means of averting war." Only with tactical atomic weapons in the country's arsenal would "the defense of Europe with reasonable numbers of men" become "a practicable matter."[62] He also advocated a retreat from strategic bombing and an embrace of a tactical A-bomb strategy several months before Robert Oppenheimer went out to Pasadena to join the conferees on Project Vista, where he helped draft such ideas into the chapter of the project report that dealt with atomic weapons. Other evidence suggests that Bush may have exerted a significant influence on Oppenheimer's thinking. The famous phrase about "two scorpions in a bottle," which Oppenheimer used in his 1953 *Foreign Affairs* piece, was borrowed from Bush, who used the analogy in a talk at Princeton.[63]

Throughout the early 1950s he took advantage of his friendship with Omar Bradley to try to get his ideas understood. Between 1950 and 1952 he peppered Bradley with letters expressing his uneasiness about a range of matters, especially about the growing irrelevance of scientists as advisers and consultants. In February 1950, he told the general that he was "appalled" at the military's excessive reliance on airpower, and he noted that another war "is not going to be won by the Air Force alone, but by the Army, Navy and Air Force in collaboration and concert." The air force had "to get down to earth" and to "substantiate its programs before a tribunal competent and willing to judge them from every angle on a ruthlessly analytical and factual basis." This need applied to all of the services, but "particularly to the Air Force."[64]

The reckless and wasteful spending by the Pentagon was all the more outrageous since it was bringing so few results. "What," Bush asked Bradley, "have we done with the three billion dollars" that had been spent on military research since 1945? The war in Korea, he argued, revealed a striking unwillingness or inability on the part of the military to introduce new weapons and techniques or even to use the innovations from the last war. How was this money being spent and, apparently, being wasted? No one was in a position to make any honest determinations because the military resisted any civilian intrusions into its doings.[65]

Bush was eager to assure Bradley that in spite of his criticisms he was writing as a friend of the military. "On the whole . . . our military establishment functions rather well," he wrote. "At least it is permeated with healthy personal relationships, and there is genuine mutual respect between men in uniform and professional civilians of various sorts." But, Bush warned, "there is something wrong." To him it was clear that the "something" was the absence of any real check on the way the services went about their spending and procurement plans without any oversight or input from reasonably disinterested parties.[66]

Bush realized that this problem originated in part from the very success of scientists in providing the military with new weapons. He might have added that decades of corporate and academic propaganda about the wonders of scientific innovation and evolution distorted the ways in which the military thought about weapons development. The generals and the admirals, Bush complained, looked on the scientists and the engineers as "magicians" who would "come around and pull a rabbit out of a hat," leaving the military people to "decide whether it is a good rabbit." But this was not the kind of professional partnership he envisioned in the late 1930s and early 1940s. "Sitting around in a receptive frame of mind to witness miracles," Bush complained to Bradley, "isn't going to get anywhere." The planning in the Pentagon had to be more efficient and more realistic. Scientists had to be involved in that planning. Bush worried to Bradley about how to "slam that idea in until I make a dent somewhere . . . without weakening the public confidence" in the military.[67]

The only thing being slammed was Omar Bradley's mailbox, as Bush continued to pressure the Joint Chiefs chairman to adopt his suggestions for reform. He even took the occasion of the publication of Bradley's memoirs both to congratulate the new author and to "deplore" the way in which he had slighted the accomplishments of the OSRD, particularly with regard to the proximity fuse. Bush could not help noting that Bradley, as well as Eisenhower, Marshall, and "you chaps at the top" generally, had given little attention to the importance of the innovations provided by the OSRD. To Bush this oversight was indicative of the "blind spot" that continued to exist in the military regarding new weapons and the importance of using technical expertise.[68]

The military certainly did not have a blind spot when it came to the Super; Bradley was all for it. Here again Bush's position seemed contradictory. It was hard to be a zealous advocate for the exploitation of scientific innovation while simultaneously calling for restraint with regard to the most spectacular innovation the scientists had been able to come up with. But Bush, of course, made a distinction between different weapons. He did not have much regard for the massive Super as a battlefield weapon. He called for the implementation of procedures for getting smaller, novel technologies out of the laboratory and onto the battlefield in rapid fashion. It was smaller technologies like the wartime proximity fuse which Bush believed the military needed to develop, not unusable weapons like the gargantuan Super.[69]

But even when Bush did begin to go public with his criticisms, he was determined not to cause offense at the Pentagon. In November 1951, he told Bradley he had written an article that might be published in the *Saturday Evening Post*. "Perhaps instead of putting this article before the public," he offered, "it ought to be put before the Defense establishment; after all that is the group it is aimed at." Then, if the memo "internally produced some discussions resulting in constructive effort, it might put me in a position to write a very different article boosting something that is under way instead of calling for something that seems to be absent."[70] In September 1952 he alerted Bradley to the criticisms he was about to level in a speech to be given at the Mayo Clinic. "I am going to use great care," he assured the general, "not to be thought critical of your colleagues on the Joint Chiefs of Staff and most certainly not of you personally."[71]

This desire not to offend and not to appear to be taking an adversarial position inevitably led Bush to limit his criticisms entirely to organizational and procedural matters. In a December 1952 piece in *Collier's* titled "What's Wrong at the Pentagon?" he argued that the organization of the country's military establishment was flawed. It was not the leaders themselves who were at fault. The problem was that the military leaders "are placed in an unworkable position. The system under which they function is wrong at the top. The gears grind. With sound planning impossible at the top, we inevitably miss opportunities for improving our defensive arrangements. The result is frustration and confusion."[72]

The ideas of both Bush and Oppenheimer were becoming increasingly unacceptable in the political climate that existed in the early 1950s. By 1952, proposals to limit the nuclear arsenal or check the powers of the military chiefs found no sympathetic ears in the military, in the executive branch, or even in the Congress. Brien McMahon, the onetime advocate of civilian control, had long since fallen into the grip of nuclear fanaticism. A second and a third Soviet atomic test was held in the fall of 1951, and that same September McMahon gave a speech in which he called for the nuclearization of all three services. He wanted the United States to "go all-out in atomic development and production," and he wanted the nations of the world to join the United States in a "moral crusade for peace and freedom." When he heard rumors that the Mike shot might be postponed, he threatened to have Truman impeached if he blocked the test. He also, tragically, fell ill from cancer in 1952 and would be dead by July. Near the end of his life he telegrammed his state convention, which was considering drafting him as a presidential or vice-presidential candidate, that if elected he would order the AEC to produce hydrogen bombs by the thousands.[73]

Bush continued to pursue his advocacy of internal, organizational changes. He was a member of Nelson Rockefeller's committee on the reorganization of the Department of Defense, which submitted a report in April 1953 that called for the elimination of the Research and Development Board, the creation of an under secretary of defense for research and development, the diminution of the power of the service heads, with increased authority going to the defense secretary, and the alteration of the Joint Chiefs' role from that of commanders to advisers.[74]

Walter Millis has aptly characterized these proposals as an attempt to model the military bureaucracy "upon the example of the giant modern industrial corporation." The secretary of defense was considered the equivalent of the CEO; the service secretaries were the vice-presidents in charge of the operating divisions; the assistant secretaries were vice-presidents for functional areas; and the Joint Chiefs were the long-range planners. The reforms were enacted, but they failed to cure the problems related to procurement and waste. The attempt to construct a better business model failed because the military was not a corporation but a unique bureaucracy in which political problems were indistinguishable from military ones.[75]

While Bush remained mired in his efforts to attain some organizational solution, Oppenheimer realized that the fundamental political premises of the arms race and of the Cold War had to be challenged. He decided that if the "Armaments and American Policy" report could not be disclosed, then he would simply express the ideas it contained in a modified manner in some public forum. With the consent of Eisenhower, and the horrified disapproval of Strauss, Oppenheimer made a speech before the Council on Foreign Relations in New York which set forth the nonclassified aspects of his argument. The speech was later published in the council's journal, *Foreign Affairs*, and also reprinted in 1955 in a collection of essays titled, perhaps sardonically in light of the events of 1954, *The Open Mind*.

He broadened his calls for greater candor into a general indictment of the culture of secrecy that dominated nuclear weapons policy. "We do not operate well," Oppenheimer wrote, "when the important facts, the essential conditions, which limit and determine our choice, are unknown. We do not operate well when they are known, in secrecy and in fear, only to a few men." In an obvious reference to Truman, Oppenheimer cited the example of an "ex-President" who had expressed doubt about whether the Russians did, in fact, have nuclear weapons. Such a preposterous comment, coming from someone "who has been briefed on what we know about the Soviet atomic capability," could be made only amid a culture of secrecy. "Perhaps . . . it was all so secret that it could not be talked about, or thought about, or understood. . . . Such follies can occur only when the men who know the facts can find no one to talk to about them, when the facts are too secret for discussion, and thus for thought."[76]

The culture of secrecy also shut out any possibility of informed public opinion being brought to bear on nuclear questions. "Public opinion cannot exist in this field. No responsible person will hazard an opinion in a field where he believes that there is somebody else who knows the truth, and where he believes that he does not know it." Here Oppenheimer was re-opening the questions raised by the Franck Report and by Lilienthal, but he did so while also trying to make a distinction between military secrets, which were justifiably preserved, and political secrets. "Knowledge of the characteristics and probable effects of our atomic weapons," Oppenheimer wrote, could be disseminated without causing injury to America's national security. "The numbers available," he continued, "and . . . the changes that are likely to occur within the next years, this is not among the things to be kept secret. Nor is our general estimate of where the enemy stands."[77]

To Strauss these proposals were anathema. The suggestions of Oppenheimer, he noted in his memoirs, would have greatly facilitated the Soviets' espionage efforts. Strauss may have considered that to be Oppenheimer's precise intention. Indeed, in the Q&A session that took place at the Council on Foreign Relations after delivering his address, Oppenheimer stated, "My own view is that the enemy has most of the information and we do not. . . . Secondly, I think it would be good if the enemy knew it, very good." Given the inflamed and suspicious atmosphere around Oppenheimer by 1953, that comment was likely to be interpreted in the most cynical light. Strauss's reaction to these ideas is indicated in a note he sent to C. D. Jackson in August 1953, in the aftermath of the announcement that the Soviets had tested a purported thermonuclear device. Strauss underlined for Jackson the portions of Oppenheimer's *Foreign Affairs* piece in which he had said he believed the Soviets were four years behind the United States in thermonuclear matters.[78]

But Eisenhower liked what he heard. He set C. D. Jackson the task of drafting a speech which by the time it was delivered had evolved away from the principles of candor called for by Oppenheimer and had become instead the "Atoms for Peace" proposal of December 1953. Eisenhower did, however, share some of Oppenheimer's ideas about the need for educating the people about the dangers of the arms race. Unless there was such education, "we will drift aimlessly, probably to our own eventual destruction. . . . Among other things, we should describe the capabilities now and in the near future of the H-bomb, supplemented by the A-bomb." He often remarked to his aides, "We've just got to let the American people know how terrible this thing is."[79]

His agreement with Oppenheimer was most vividly expressed in his "Chance for Peace" address, which he gave on April 16, 1953. Some of the most eloquent passages were drafted by the fire-eating author of NSC-68, Paul Nitze. "The worst to be feared and the best to be expected can simply be stated. The worst is atomic war. The best would be this: a life of perpetual fear and tension; a burden of arms draining the wealth and the labor of all peoples. . . . Every gun that is made, every warship launched, every rocket fired, signifies, in the final sense, a theft from those who hunger and are not fed, those who are cold and are not clothed."[80]

In the whole dreary history of the Cold War, it is hard to find a person more eloquent against the dangers of war than Dwight Eisenhower. Ironically and tragically, he allowed his ideas to be overridden by the bureaucracy that Lewis Strauss now controlled. Presidential secretary Ann Whitman revealed the extent to which Eisenhower relied on Strauss and left the business of supervising atomic energy matters almost exclusively to him. "As far as I can judge," Whitman wrote in her diary, "the President has only to make the top decisions in this field, and they take a minimum amount of time." This was in contrast to Eisenhower's much more active involvement in other cabinet areas.[81]

Strauss was able to mount more energetic and devastating assaults against Oppenheimer once he returned to the inner circle of state science and accepted the directorship of the AEC in the spring of 1953. He leaked information to a *Fortune* magazine journalist, who did a hatchet job on Oppenheimer titled "The Hidden Struggle for the H-Bomb: The Story of Dr. Oppenheimer's Persistent Campaign to Reverse U.S. Military Strategy." Strauss had even made his very acceptance of the new post conditional upon Eisenhower's promise that Oppenheimer would have nothing to do with the organization. In July 1953, five days after officially taking over, he had Oppenheimer's access to classified documents restricted, and he ordered AEC officials to seize the documents maintained by Oppenheimer at Princeton and had them placed in a new repository directly under AEC control.[82]

In December 1953, Strauss received a copy of William Borden's hysterical letter to the FBI accusing Oppenheimer of being "more probably than not" a Soviet agent. Strauss distributed the letter to the other commissioners. Secretary of Defense Charles Wilson and National Security Adviser Robert Cutler discussed it with Eisenhower. Thinking of how Republicans had attacked the Truman administration for maintaining Harry Dexter White in its employ after White had been accused of being a Communist, Wilson and Cutler stressed the political damage Oppenheimer's case might cause the administration.

Cutler broached the idea of approaching Vannevar Bush since he was a "friendly scientist" and asking him to intercede discreetly with Oppenheimer and convince him that the best course would be to resign his consultancy. There is no evidence of what, if any, actions Bush took at this stage. No quiet settlement was reached and Eisenhower suspended Oppenheimer's clearance, unwittingly setting in motion a chain of events that would remove from the government the most prominent scientist whose views most closely coincided with his own.[83]

CHAPTER 6

---

THE OPPENHEIMER CASE, EISENHOWER, AND
THE TRIUMPH OF SCIENTIFIC MILITARISM

---

The record of Oppenheimer's security hearing provides fascinating documentary evidence of what Vannevar Bush would later describe as the "thought control" demanded by the Cold Warriors who dominated state science. It represented a vivid and disturbing statement of the marginal, subservient role the political and military leaders of the state expected from the scientists. It also demonstrated the degree to which Oppenheimer's supporters, and even Oppenheimer himself, were willing to subscribe to the assumptions of the inquiry, namely, that there was something decidedly wrong with opposing the Super, that there was something wrong with being a scientist with strong views on broad political matters, or at least those that challenged the Cold War creed, and that it was wrong even to question the idea that the United States should have the "strongest possible offensive capability." In their readiness to appease their attackers and in their unwillingness to fight publicly against the gross injustice being carried out against Oppenheimer, the leading scientists left themselves no viable ground in the public realm from which to resist the triumph of scientific militarism.

The extent of that triumph can be seen in the ways that even President Dwight Eisenhower accommodated himself to the ideas of scientific militarism in spite of his instinctive uneasiness over its implications. But Eisenhower, like the moderate managers of state science, embraced the paralyzing notion that corporate prosperity and the proliferation of weapons technology were two sides of the same ideological coin. Both were the results of the evolution of American science and American enterprise and thus were beyond political challenge. He saw his role as simply one of managing the balance between economic and military strength. Not surprisingly, then, he deferred to Lewis Strauss on matters of science and weapons policy. Strauss was a successful businessman and a hawkish anticommunist. He seemed the perfect champion of American capitalism as well as American militarism. So when Strauss launched his attack on Oppenheimer, Eisenhower chose to stand aside and just let it happen.

In some ways, so did Robert Oppenheimer himself. His response to the charges against him was fundamentally a defensive one: yes, he argued, I am a hawk on American weapons policy. His supporters tended also to stress his toughness, his hardheadedness, and his anticommunism. Conant, for example, underscored how Oppenheimer was really "hard headed, realistic, and thoroughly anti-Soviet." No one, with the notable exception of Vannevar Bush and I. I. Rabi, challenged the very nature of the proceedings and denounced them for what they were: a trial for the expression of dissenting political beliefs.[1]

Oppenheimer, in his response to the AEC's letter setting forth the list of charges, took a reflective, discursive approach that was very much in character but ill suited for the political combat he was about to join. "The items of so-called derogatory information set forth in your letter," he wrote, "cannot be fairly understood except in the context of my life and my work." That was a defensive response, which conceded that his life required explanation and placing in context. In his opening presentation to the Gray Board, Oppenheimer stressed his credentials as a hawk. He pointed out how the GAC not only approved but also sometimes initiated measures that increased the strength of the military. This was particularly so in the earliest days of the commission, when the members of the GAC were more knowledgeable than the rookie commissioners. The commission, Oppenheimer recalled, "relied on us very heavily," and the advisers led their bosses to an understanding that "the principal job of the Commission was to provide atomic weapons and good atomic weapons and many atomic weapons." The so-called peaceful atom was almost an irrelevance. The AEC and the GAC were concerned primarily with strengthening the country's nuclear arsenal. The desire to strengthen the country's military power even explained the advisory committee's controversial recommendation against the Super. Oppenheimer emphasized that the decision reflected the conviction that "such a program might weaken rather than strengthen the position of the United States."[2]

But in the face of a vicious cross-examination that was strengthened by the illegal wiretaps Strauss had placed in the offices of Oppenheimer's attorneys, Oppenheimer confessed to having lied to Manhattan Project security officials over the infamous Chevalier incident. Paradoxically, Oppenheimer may have been telling the truth to officials back in 1943 and perjuring himself in 1954. That tangled tale has been extensively addressed by others and is not relevant to the issues I want to address in this chapter. The point of the whole exercise for Lewis Strauss was to destroy Robert Oppenheimer. And it succeeded.[3]

Near the end of the hearings, the chief interrogator, Roger Robb, asked

Oppenheimer with what seemed to be genuine perplexity, "why you felt it was your function as a scientist to express views on military strategy and tactics." Oppenheimer explained that "having played an active part in promoting a revolution in warfare, I needed to be as responsible as I could with regard to what came of this revolution." Asked if he favored the Super in 1954, Oppenheimer replied with a bitter wryness that was probably lost on his interrogators: "I believe everybody today is an enthusiastic proponent."[4] Such "rehabilitation" did nothing to sway the Gray Board in Oppenheimer's favor. While the majority of the board found "no evidence of disloyalty," it nevertheless revoked Oppenheimer's clearance. In spite of "much responsible and positive evidence of the loyalty and love of country of the individual concerned . . . we do not believe that it has been demonstrated that Dr. Oppenheimer has been blameless in the matter of conduct, character, and association."[5]

The board was determined to adopt the much looser standards of Eisenhower's 1953 Executive Order, which allowed them to find that "any doubts whatsoever must be resolved in favor of the national security." And there were plenty of doubts, all of them centering around Oppenheimer's position on the Super. "The Board finds, that if Dr. Oppenheimer had enthusiastically supported the thermonuclear program either before or after the determination of national policy, the H-bomb project would have been pursued with considerably more vigor, thus increasing the possibility of earlier success in this field."[6]

The board also found that Oppenheimer had demonstrated insufficient loyalty to the security system itself. "There remains also an aspect of the security system," the majority opinion read, "which perhaps has had insufficient public attention. This is the protection and support of the entire system itself." Scientists had to exhibit "a wholehearted commitment to the preservation of the security system and the avoidance of conduct tending to confuse or obstruct." The means as well as the ends of state policy were now to be treated with reverent respect.[7]

The board seemed intent on making a statement about the proper place of scientists within the high echelons of American politics. "These specialists have an exponential amplification of influence which is vastly greater than that of the individual citizen," and though the board recognized that it was "vitally important that Government and scientists alike understand the need for and value of the advice of competent technicians," they advised that "caution must be expressed with respect to judgments which go beyond areas of special and particular competence." In the minds of the board members, there was only one kind of advice a "competent technician" could give, and they urged "Government officials" to be sure they got nothing less: "In evaluating advice from a specialist which departs from the area of his specialty, Government officials charged with the military posture of our country must also be certain that underlying any advice is a genuine conviction that this country cannot in the interest of security have less than the strongest possible offensive capabilities in a time of national danger."[8]

On this fundamental point the board found Oppenheimer wanting. "We cannot dismiss the matter of Dr. Oppenheimer's relationship to the development of the hydrogen bomb simply with the finding that his conduct was not motivated by disloyalty, because it is our conclusion that, whatever the motivation, the security interests of the United States were affected." The board was especially concerned that "he may have departed his role as scientific adviser to exercise highly persuasive influence in matters in which his convictions were not necessarily a reflection of technical judgment, and also not necessarily related to the protection of the strongest offensive military interests of the country." But then, astoundingly, the following sentence appears in the opinion: "In any event, the Board wishes strongly to record its profound and positive view that no man should be tried for the expression of his opinion."[9]

The Gray Board having abandoned not only its original charges but logic and intellectual honesty as well, the AEC commissioners wrote opinions that reflected the political free-fire zone that the board had created. Since the question of loyalty was now only one of several factors to consider in a security investigation, the commissioners came up with unique and individualized justifications for their decisions against Oppenheimer. Kenneth Nichols stressed Oppenheimer's "dishonesty." Strauss deplored his scandalous "associations." Thomas Murray argued that Oppenheimer really was "disloyal" according to the standards advanced in the Gray Board's findings: he was not "loyal" to the rules of the security system.[10]

All of these arguments were simply flimsy covers for the true purpose of the proceedings. The attack on Oppenheimer was an attack on the particular brand of scientific leadership he represented, one that was willing to advance alternative visions concerning American weapons policy and the public role of scientists. Oppenheimer, Bush, and Conant had been trying to wrest the question of nuclear weapons out of the intellectual and cultural context into which it had fallen in the late 1940s and early 1950s. None of these three men, after 1950, regarded nuclear energy as a panacea for civilian society. None of them saw it as an unalloyed virtue and just another admirable expression of Yankee ingenuity, and all of them perceived its dangers. All of them came to adopt something of Jewett's caution and fear about the destructive power of science; they also perceived the distorting effect the weapons would have on military planning and on political leadership. But nevertheless, they could not bring themselves to adopt an openly adversarial position against this new system of scientific militarism.

These men were not radicals and they were too accustomed to being powerful, highly respected insiders. Their entire professional identity hinged on their associations with the leaders of the nation's political and military elite. The moderate state scientists seemed to lack any language or any coherent set of principles that they could publicly assert against the shameless red-baiting and coercion taking place before their eyes.

Conant, to be sure, showed admirable courage in testifying on behalf of Oppenheimer in spite of the resistance and the subtle threats of his boss, Secretary of State John Foster Dulles. Dulles told Conant, who at the time was the U.S. high commissioner for Germany, that an appearance could destroy his usefulness in the government. "I said I quite realized this," Conant recalled, "and he only had to say the word and I was through!" But he did not make his outrage public knowledge. "The atmosphere in the U.S.A. today is pretty close to a mild reign of terror," Conant wrote in his diary. But he consoled himself with the hope that "if the leading citizens get upset about this business as they are beginning to, I think the tide will turn." The system would always, somehow, work itself out of trouble through the intercession of the insiders.[11]

Bush was also profoundly outraged by what was happening to Oppenheimer, but he vented none of his fury publicly at Strauss, whom he surely knew was the author of Oppenheimer's show trial. Again he directed his criticisms toward the organization of the system. The "security system" had to be improved, he argued. But he would not "name names" and attack the people he knew to be perverting the system. In the spring of 1954, he told Ernest Lawrence's right-hand man, Don Cooksey, that he could "only hope that the whole thing may gradually stabilize into something that is more wholesome and perhaps it will. In the meantime, all we can do is make the best of some of the minor irritations and try to steer the general trend a bit here and there as we find opportunity."[12]

Bush declined to apply direct pressure to areas where it might have been to good effect. During the first week of the Oppenheimer hearings he received out of the blue a letter from Lewis Strauss in which the admiral wrote, "Last night Don Quarles told me of your talk with him and the mention of me. *Please* reserve judgment until I am free to speak with you on the subject which should be in two or three weeks after the Board has reported." Bush responded by expressing surprise at the implication that he had been discussing Strauss with Quarles. "The fine relations I have had with you for many years," Bush assured Strauss, "have not altered in any way whatever. My whole thought, in these trying days, is to attempt to further effective relations between scientists and their government, which I labored to create during the war years, and which are now in jeopardy."[13]

Bush was talking to the man who was chiefly responsible for destroying the institutional alliances he had built up over the preceding fifteen years, and the ensuing passages of his letter read like a threat. "I have thought of writing the President, for it seems to me that he only can speak the words which will put the attitude of government in proper light." But, Bush continued, "I have not sent a letter yet. It seems to me that, if I point out the problem to the President, I am bound to suggest a way in which he could act to solve it, and this has me baffled. But disintegration of morale is becoming cumulative, and some prompt action to turn the tide seems to me essential. The difficulty, of course, is to do this without interference with orderly procedures. So I have hesitated. But I am deeply troubled."[14] The threat was real, but so too was Bush's hesitancy. No evidence has surfaced indicating that such a letter was sent to Eisenhower.

In late April, Bush wrote Strauss again. He said it was essential for Eisenhower to speak out against "thought control." It was "not a matter of politics, or even of balance of powers between Legislative and Executive branches. It is a matter of principle and decency." Bush seemed determined not to see political issues whenever he could. He closed his letter with the impotent comment, "I wish I knew how to help."[15]

When he testified before the Gray Board his temper exploded, and his words clearly expressed his sense that the attack on Oppenheimer was also, indirectly, an assault on him and his notion of how scientists should work with the state. He directed his wrath especially at the charges relating to the hydrogen bomb, which were, in his words, "quite capable of being interpreted as placing a man on trial because he held opinions, which is quite contrary to the American system, which is a terrible thing." He reported that in his travels he found people thinking the same way he did: "that here is a man who is being pilloried because he had strong opinions, and had the temerity to express them." No board, Bush said, "should ever sit on a question in this country of whether a man should serve his country or not because he expressed strong opinions. If you want to try that case, you can try me. I have expressed strong opinions many times, and I intend to do so. They have been unpopular opinions at times. When a man is pilloried for doing that, this country is in a severe state."[16]

Bush became even more outspoken in the aftermath of the Oppenheimer affair. In June 1954, he published an angry piece in the *New York Times Sunday Magazine*, "If We Alienate Our Scientists," in which he lamented the collapse of the alliance he worked so hard to build up. "During the war," he wrote, "there was produced an effective working partnership between scientists and military men, in fact, more broadly, between professional men and officers of Government." They learned to work together under terms of mutual respect. But now, the partnership was "gravely damaged" and was being "gradually destroyed."[17]

A large part of this disintegration, Bush noted, was the result of the runaway "security system" that had been installed in the government, and he once again used the term "thought control" to denounce that system. But he realized something else which the findings of the Gray Board and the majority of the AEC made clear in the Oppenheimer case. Behind the hysterical witch-hunting for reds there was another motive that drove this persecution of scientists: a reaction against the idea that scientists could play a prominent role as public experts. "The scientist," Bush argued, "is not beyond his field in considering modern military strategy and tactics. Every such question has its military and also its scientific aspects, and only by the fusion of these can it be analytically analyzed." Scientists feel they are being pushed out, Bush noted, "and indeed they are." He bitterly resented this diminution of the status and prestige of his profession. "Scientists need to be used not as lackeys or underlings but as partners in a great endeavor to preserve our freedoms," he argued. But the partnership was breaking up, and Bush was becoming more vehement in his criticisms.[18]

At a December 1954 meeting of the American Association for the Advancement of Science in Berkeley he said that the threat to American freedom was greater at home than from abroad. In March 1955, he wrote another angry piece for the *Times* Sunday magazine, "To Make Our Security System Secure," in which he again denounced the "thought control" that was being practiced in the name of weeding out subversives. Bush aimed his vehement objections at the security system but not at the Cold War ideology, of which the security apparatus was only a symptom. Even when he used extreme terms like "thought control," he could not bring himself to embrace the bold political measures required to eradicate such a problem.[19]

Bush may have felt some indirect retaliation for his defense of Oppenheimer before the Gray Board. His longtime friend and his former deputy at both MIT and the OSRD, Carroll Wilson, found himself running afoul of Strauss's fanatical security regime in the summer of 1954. Wilson had changed jobs in June, moving from the Climax Melybnedum Company to the Metals and Control Corporation, of which Bush was a director. As vice-president and general manager at his new firm Wilson would oversee some classified projects the company had on contract with the AEC. It was routine procedure to recertify one's clearance upon changing jobs, and Wilson, as a former AEC general manager himself, knew that this was a process that usually took days or, at most, weeks to carry out. In Wilson's case it was taking months, not only causing him embarrassment at his new firm but threatening his very usefulness as a supervisor.

Wilson turned to Bush for help, explaining to him in August 1954 that his clearance was being held up "for reasons you and I very well understand." But the clearance had still not come through by early November. Finally, a meeting was held between Wilson and commission representatives. A "friendly" discussion ensued and Wilson discovered that the events that had stalled his clearance were "old chestnuts" that had occurred between 1947 and 1950 in "which there was dissent on the part of one member of the Commission. The correlation was almost 100%."[20]

The dissenting commissioner was an unmistakable reference to Strauss. Bush could not possibly have failed to understand Strauss's central role in this "shabby episode," as Wilson called it, nor could he have been in doubt about the admiral's role in the far more sordid and illegal effort against Oppenheimer. But he refused to take on Strauss publicly. In fact, Bush had been "appalled" when Wilson publicly declared his reasons for resigning from the AEC as general manager back in 1951; in his resignation letter to Truman, Wilson explained that he could no longer serve under chairman Gordon Dean, whom he believed was appointed purely for political reasons at the behest of Dean's former law partner, Senator Brien McMahon. Such public airing of insiders' disputes horrified Bush.[21]

Lilienthal also displayed a similar tentativeness in the face of the attacks against Oppenheimer, as well as himself. He thought of publishing an opinion piece somewhere to defend Oppenheimer—and his own record. He wanted to take issue with the hysterical thinking that led to the investigation. But, as he recorded in his diary, he thought back to the indifference that greeted his October 1953 *New York Times Sunday Magazine* piece about the H-bomb decision. The public was indifferent, he thought, because the piece lacked backstage gossip or an emphasis on personalities. "And I don't want to do anything now," he wrote in April 10, 1954, "to add fuel to the flames of feeling, excitement, and fear. So I had better contain myself."[22] When preparing to testify at Oppenheimer's hearing, Lilienthal was apprehensive and vigilant. He knew Lewis Strauss all too well, and he knew that this was not just about Oppenheimer but about the whole moderate approach to the management of nuclear weapons which he, Oppenheimer, Bush, and Conant had advocated.[23]

Significantly, Lilienthal recorded a conversation he had with Andre Meyer, whom he referred to as his "boss." Meyer, a partner in the Lazard Frères investment banking house, was a major source of investment capital for Lilienthal's Third World development projects. After the drubbing Lilienthal received at the hands of Roger Robb before the Gray Board, Meyer consoled him by saying that "decency in Government is almost gone, for the time being perhaps. But we want to show you that there is still a great deal of decency and fairness and honor left among businessmen in private life, a great deal of it."[24]

Lilienthal had come a long way from the New Deal idealism of his Norris days. He was now a comfortable resident of Sutton Place, working cheek by jowl with the "Wall Streeters" he had denounced as partners in the effort for international control. In 1953, Lilienthal published a book, *Big Business*, which celebrated what he perceived as the new responsibility of the corporate elite and the social efficiency of large-scale economic concentration. The book represented a businessman's elaboration of the ideas expressed in the TVA book. He also abandoned the utopian notions of the peaceful atom, which he had somewhat recklessly spouted during his early period on the AEC. By 1954 he felt that talk about inexpensive nuclear power, such as that emanating from the mouth of Strauss and the publicity department of GE, was "hopelessly optimistic and unrealistic."[25] Lilienthal was not the only one who seemed to be trimming his sails in the face of the harsh, prevailing political winds. Even Arthur Compton, who in 1955 took a brave stand in defense of the blacklisted Berkeley scientist Martin Kamen during the latter's ordeal at the hands of the congressional red-baiters and Colonel McCormick's *Chicago Tribune*, produced a memoir in 1956 that seemed to legitimate the persecution of Oppenheimer: "Some of Oppie's friends are, or have been, on the Communist side of the line. This has made it necessary for those in positions of highest responsibility to reluctantly withdraw from him access to our military secrets."[26] The basic beliefs of the moderate state scientists still militated strongly against a politicized interpretation of their public role. They advanced an image of themselves as equal but unexceptional partners in the governing elite, and they continually sought to ratify the larger political system that provided them so many rewards, even when that system was viciously turned against their friends and colleagues—or even themselves. Even when they admirably criticized the arms race or courageously criticized the security system, they managed to legitimate enough of the system's underlying ideas to ensure that it would continue to function and thrive in spite of their occasional protests.

Bush perceived some of this after the Oppenheimer debacle. "Ever since the first bomb went off at Almagordo [sic]," he wrote Conant, "there have been many of us who have envisioned quite clearly what it would lead to. Now after ten years, we face a situation which will soon be developed to its full threat. We and Russia will face each other over piles of bombs, and it is highly probable that both will have means for delivering such bombs on targets." The situation, Bush felt, was "an appalling one." The architects of the American arsenal "emphasized the desirability of a striking force, without equally emphasizing the other side of the shield." Bush ruminated over his career as failed arms control advocate, and he expressed some uneasiness, if not guilt. As part of the 1952 disarmament panel, he "advocated increasing the military strength of the United States as rapidly and as completely as possible," yet he also noted that "we . . . searched assiduously for a way out of the morass." He recalled his futile effort to get Acheson to stop the 1952 test and noted that "the entire program for the building of an H-Bomb was so vigorously under way that any suggestion of delay received practically no consideration whatever."[27]

On the eve of the Oppenheimer affair he ruminated to his friend Don Price that "the problem of how far a technical man working with the military is entitled to speak out publicly is quite a question. . . . I kept in channels rather religiously, perhaps too much so." There was a fundamental tension that Bush could not adequately resolve in his own mind. On the one hand, he believed strongly that a person who participated in internal government debates "had better not talk about it publicly . . . for he is likely to destroy his usefulness." But on the other hand, scientists "are also citizens," and "when an individual citizen sees his country going down a path which he thinks is likely to be disastrous he has some obligation to speak out."[28] The essence of this "quandary," as Bush called it, hinged on the meaning of the term "usefulness." Bush and Conant, even in their most outspoken and critical moments, were still intent on making themselves "useful" to the system of state science which they had done so much to create and which had provided them with the excitement and prestige of moving in and around the center of national power. It was practically impossible for them fully to break away and become adversarial critics of that system.

The story of scientific politics in the 1950s makes it clear that there were a variety of factors driving events: ideology, national politics, institutional agendas, competing scientific cultures, and petty personal vendettas. One cannot easily imagine Robert Oppenheimer being pilloried if someone other than Lewis Strauss had been head of the AEC in 1954 and someone other than Edward Teller had not had his views embraced by the military and political elite. The institutional and the ideological momentum of the Cold War—which these men helped create and accelerate—was immensely powerful in those years, and it simply overwhelmed the moderates.

But the moderates had been instrumental in creating the very ideology that was, in modified form, being used against them. One cannot imagine the momentum of scientific militarism becoming so dominant in the first place if the scientific administrators had not subscribed to the creed of corporate science, with its stress on the politically redemptive powers of innovation and technological evolution and the self-evident virtues of elite collaboration. It was not momentum, per se, that determined the triumph of scientific militarism. It was the historical choices of a generation of scientific administrators who eschewed open political conflict and who based their authority on collaborative organizational patterns that made it possible for a Lewis Strauss and an Edward Teller to use that institutional structure for their own purposes without any effective opposition from insiders.

To explain the defeat of moderates like Bush, Conant, and Lilienthal simply by invoking the Cold War, or McCarthyite hysteria, or even the rise of a repressive "power elite" is insufficient. It ignores their own intellectual responsibility for creating the ideological and institutional climate that did them in, and, more important, it ignores the historical fact that there *was* opposition to this system and the ideologies that sustained it, but from outsiders.

The 1950s also saw the birth of an assertive disarmament movement. Even when Bush and Conant argued against developing the Super, they were still far away from the ideological principles of that movement. Groups like Scientists against Nuclear Energy (SANE) stressed the health dangers of nuclear radiation and the importance of world peace. They were not primarily concerned with strengthening the institutions of the corporate economy or the American state. That difference in priority made all the difference in the world in terms of their goals and the methods they used to pursue them. The Benjamin Spocks, the Ralph Lapps, and the Josef Rotblats of the world could afford to flout the good opinion of the powerful and the "insiders" because they were in rebellion against those very people. Their story is not the one being told in this book. But there were possible alternative paths that Bush, Conant, and even Lilienthal could have tread in the 1950s. To explain the path they did take in terms of ideological context, institutional momentum, and professional cultures is hardly to excuse them for failing to choose different ideologies, different professional cultures, and a different kind of momentum. Nor is it to suggest that they could have changed their political and cultural context easily. But they *were* in a position to change it: they were prominent, respected, and even famous public figures who could have commanded an audience before any of the major media, or even the president of the United States, to speak out against what they all knew Lewis Strauss was up to. They chose not to do so.[29]

It would be Eisenhower himself, ironically, who would slowly and belatedly challenge Strauss's single-minded hawkishness, but by the time he did so in the late 1950s, the ideological and institutional base of scientific militarism had solidified. In 1954, Dwight Eisenhower presided over a military academic industrial complex that was expanding dramatically in terms of money, bureaucracy, and, most significantly, ideological power. The Defense Department's R&D budget climbed from $530 million in fiscal year 1949, to $510 million in FY 1950, $1.3 billion in FY 1951, and $1.6 billion in FY 1952. By late 1951 Defense Department and AEC contracts constituted 40 percent of the money devoted to industrial and academic research.[30]

Once Lewis Strauss closed his fist around the powers of the AEC in 1953 he placed in key positions only those scientists who advocated the "strongest offensive capability" for the United States, men such as Teller, Lawrence, and John Von Neumann. These advisers used the ideology of unrestrained scientific innovation in the form of arms testing and proliferation to advance an extremely militarized vision of the state.

But by his second term, Eisenhower found this vision unacceptable and even intolerable. He turned to other scientists who did not share Strauss's fanaticism and began belatedly to control the arms race and the whole culture of scientific militarism. Tragically, he began that process too late and too weakly to be truly effective. Cold War politics and scientific militarism had taken such strong root in American culture and even in Eisenhower's mind that it was difficult fully to undo the damage that had already been done.[31]

Scientific militarism was in full stride when Eisenhower took over in 1953. At the same time that the battle against Oppenheimer and the moderate state scientists raged within the government, new scientific advisory panels sprang up which solidified the subordinate and limited place for scientists in the political and military bureaucracy. These were advisory panels whose mission was limited to devising means for carrying out basic political decisions that had already been made and were to be forever beyond the critical purview of the scientists. There was no question that the armed services, especially the air force, were to have nuclear weapons. The question left for scientific advisers was simply what kinds.

Chief among the valued insiders in this period was John Von Neumann, who, according to Herbert York, was especially influential among the activist generals in the air force. York observed that Von Neumann "liked high-ranking military officers and got along very well with them. If anyone during that crucial period in the early and middle-fifties can be said to have enjoyed more 'credibility' in national defense circles than all the others, that person was surely Johnny." Generals James Doolittle and Bernard A. Schriever, the deputy chief for advance planning, told York that "it was Johnny's personal projections about the future of thermonuclear weapons, and no other individual or institutional source, that first convinced them of the new possibilities and caused the Air Force to initiate the actions that eventually led to a high-priority program to build the inter-continental ballistic missiles."[32]

In the mid-1950s, Von Neumann opened up a wide world of proliferation for the generals to consider. Nuclear weapons, he told the air force in 1954, "are no longer expensive, they are no longer scarce, and they are no longer a monopoly of the U.S. . . . In other words, one must no longer consider the nuclear component as the hardest problem involved in the weapons system of which they form part—they are now among the least difficult and most flexible parts of such systems." He also, along with Trevor Gardner, argued strenuously that the United States should devote considerably more energy and resources to the development of short-range and long-range missiles that could carry the various weapons being fabricated. Von Neumann expressed "grave concern" that the Soviets were slightly ahead of the United States in missile technology, and York believed that the intelligence of the time supported such a concern. Unless the United States made sure it had the missile technology it needed, its superiority in thermonuclear weapons would be rendered meaningless in the face of Soviet strategic missiles.[33]

Ever since the end of the war, the air force bureaucracy had been expanding around its commitment to strategic bombing with nuclear weapons. In 1950, the Air Research and Defense Command (ARDC) was created, bringing into existence a deputy chief of staff for development. But research fully blossomed only when Trevor Gardner was appointed special assistant for research and development and later assistant secretary of the air force.[34] Gardner's career in the air force and his success in bringing about the development of ballistic missiles is an especially vivid example of the way the organizational and ideological patterns established by Bush and the

OSRD were being perpetuated by newer, younger hands and for different purposes. Bush opposed a rocket program in the 1940s as both unfeasible and uneconomical.[35] But such considerations had long since stopped being prominent in the formulation of American weapons policy. Instead, the imperative of scientific and technical innovation was constantly being invoked to forge ahead with the missile program. The leaders of the program reprised the same sense of crisis and the same need for bold, technical innovation that had characterized the H-bomb controversy.

"With every tick of the clock," Trevor Gardner said in 1956, "the Soviet Union is moving closer to . . . knocking this country out. Intercontinental air power and missiles are the new double-edged sword of destruction, hanging by a hair over us all." Missile proponents such as Lieutenant General Earle E. Partridge echoed Edward Teller's attacks on the Super opponents. There would be, Partridge predicted, two schools of thought on the question of developing guided missiles. "One of these schools will be small but vigorous and will insist that the job can be done by the guided missile. The other group, representing the old fogies, will continue to insist that we adhere to the tried and proven air craft."[36]

Gardner got the 1955 budget for intercontinental ballistic missile (ICBM) research increased from $20 million to $50 million by depicting the missile program as being just as urgent as the Manhattan Project had been. He also urged air force secretary Harold Talbott to create a policy committee answerable to the president and the National Security Council. He specifically cited the precedent of Bush's policy committee, which the OSRD director had regarded as a kind of board of directors for his organization. Some Manhattan Project veterans were even brought back into service for the missile effort, including Eger V. Murphree, who became "missile czar" in February 1956.[37]

Any notion of selection or concentration of resources was abandoned when the administration decided to develop both the intercontinental ballistic missile and the intermediate-range ballistic missile (IRBM) simultaneously. This was done, in part, because of a State Department study which argued that Soviet success in achieving either of these capabilities would weaken the technological esteem in which countries of the world regarded the United States. But it also bears an echo of the Manhattan Project decision to pursue simultaneously all possible approaches to the production of fissionable uranium and plutonium. The size of the missile project ultimately came to rival that of the atomic bomb effort, while costing far more. There were two thousand contractors working on the program by the end of the 1950s, at a total cost of some $17 billion.[38]

President Eisenhower allowed the momentum of scientific militarism to continue for several reasons, some of which we have already seen. He abdicated control of the military use of science, especially with regard to nuclear weapons, to Strauss. Also, his New Look policy, with its emphasis on budget savings through troop reductions, lent greater emphasis to developing automated and electronic weaponry. These were the years of the great takeoff in the computer and high-tech industries that depended highly on military funding.[39]

But Eisenhower also had long-standing reservations about weapons spending and about the arms race in particular. He even told Lewis Strauss in 1953, "My chief concern and your first assignment is to find some new approach to the disarming of atomic energy." But Eisenhower's indifference to the many ways in which Strauss ignored and flagrantly violated this injunction only shows the laconic if not submissive attitude he adopted in response to the very limited scientific advice he received in his first term.[40]

Fortunately, Eisenhower did not submissively accept the insane advice he received with regard to waging nuclear war. He repeatedly stressed in his first term the folly of even considering such a war. As he told Korean president Syngman Rhee in 1954, "Atomic war will destroy civilization. War today is unthinkable with the weapons which we have at our command. If the Kremlin and Washington ever locked up in a war, the results are too horrible to contemplate. I can't even imagine them." Eisenhower told reporters in 1956 that atomic war would be "race suicide and nothing else."[41]

But five times between April and December 1954, Eisenhower's advisers urged him to use atomic bombs: three times in Indochina, then during the Quemoy-Matsu crisis, and once again in response to Chinese holding of American prisoners. When the NSC sent him a report suggesting the use of atomic bombs in Vietnam, Eisenhower told Robert Cutler, "I certainly do not think that the atom bomb can be used by the United States unilaterally. You boys must be crazy. We can't use those awful things against Asians for the second time in ten years. My God." Unlike his advisers, Eisenhower had already borne the responsibility of committing people to their deaths. It had been a deeply sobering experience. "If you're in the military," Eisenhower told James Hagerty in 1955, "and you know about these terrible destructive weapons, it tends to make you more pacifistic." He sensed that a similar restraint was operating in war-weary Russia. The Russians were "not ready for war and they know it. They also know if they go to war, they're going to end up losing everything they have. That also tends to make people conservative." Eisenhower explained that his experience of war and his experience writing "letters of condolence by the hundreds, by the thousands," convinced him of the importance of not being swept into war by emotions and feelings of anger or resentment.[42]

His uneasiness even extended to the matter of testing. He rejected a proposal by York—who had not yet acquired his own doubts—to test a bomb with a yield in excess of twenty megatons. None that big had ever been detonated, and authorization for this and all other nuclear tests required the president's approval. Eisenhower immediately rejected the proposal. "Absolutely not," he said, "they are already too big." Andrew Goodpaster told York that around the time of that decision Eisenhower also had commented, "The whole thing is crazy; something simply has to be done about it." James R. Killian also recalled Eisenhower remarking that "we should be giving more attention to goals of peace, such as reducing the burden of arms. After the presentation of the Gaither report, he made the same point."[43]

Eisenhower's moderate attitudes on military issues were a product of his Hooveresque philosophy of the "middle way." He was also very similar to Bush and Conant in that he had been able to rise to positions of national and international power thanks to the "organizational revolution" that had created a new "managerial class" of which he was a prominent part.[44] And like other members of this managerial class, Eisenhower both loyally supported the general political values of the institutions that fostered his career and sought to use those institutions to foster a spirit of social harmony and class cooperation.

Eisenhower's political philosophy echoed perfectly the associational ideas of Hoover and Progressives like Croly and Lippmann. "We must find a way," he told the American Academy of Political Science in 1950, "to bring big business, labor, professions, and government officials together with . . . experts and . . . study and work out these problems in the calmness of a nonpartisan . . . atmosphere." Like the Progressives and the corporate scientists, Eisenhower believed that maintaining prosperity was the central purpose of American political culture and that it could be accomplished "only when management, labor and capital work in harmony . . . ; no prosperity for one economic group is permanently possible except as all groups prosper."[45]

Like the other Hooverian Progressives we have examined, Eisenhower linked his moderation so strongly to the political interests of his benefactors that he seemed at a loss when that moderation had implications and consequences that threatened those interests. Eisenhower was supported and pushed by a faction of business leaders who also shared his ideas about a conservative, corporate commonwealth, and they correctly viewed him as the only viable Republican who could carry their ideas into power and stem the advance of New Deal "statism." Eisenhower, entirely on his own initiative, dutifully installed around him cabinet officials and advisers who were also drawn from this conservative faction of the business community, one of whom was Lewis Strauss. Eisenhower would be only the latest in a series of idealistic public servants who found his goals thwarted by Strauss's decidedly immoderate ideas and actions.

The choice of Strauss as AEC chairman was an example not only of Eisenhower's veneration for self-made millionaires but of the long-standing perception among Americans that matters of military science were closely bound up with the country's economic interests. Strauss's credentials as a successful businessman were decisive in getting him the job as atomic energy adviser. Robert Cutler explained to Sherman Adams that the post required a man who was "not a scientist or a lawyer but a practical business manager" since the AEC was "a very big business." This attitude of regarding atomic energy and nuclear weapons with the language and the mentality of business enterprise served to undercut the more sophisticated understanding that Eisenhower, as an experienced military man, brought to the problem.[46]

Strauss's "conservatism" in the 1950s was defined almost exclusively on the basis of his fear of Soviet militarism and by extension of the radicals and even liberals whom Strauss felt served as fifth columnists for Soviet aggression. Although Eisenhower never shared Strauss's apocalyptic vision of the Soviet threat, he was inclined to follow the latter's hard-line military advice based on aspects of his own experience, specifically his memories of America's lack of preparedness before World War II.

Eisenhower's centrism and his military experiences led him, while president, to err on the side of greater preparedness because the "lesson" of Pearl Harbor loomed menacingly over the American political landscape throughout the 1950s. In such a context, he considered the views of antimilitarists such as Robert A. Taft unreasonable and extreme. After one encounter with Mr. Republican, Eisenhower found himself "astonished at the demagogic nature of his tirade, because not once did he mention the security of the United States. . . . He simply wanted expenditures reduced, regardless." But it was not simply the political atmosphere that led Eisenhower to these views. His own military experience greatly influenced his thinking on these issues. Recalling the disasters suffered by the American military in the early years of World War II, Eisenhower was determined not to leave the nation militarily unprepared.[47]

But he was also determined as part of his New Look strategy to keep defense spending under control, and his former colleagues at the Pentagon and in the Joint Chiefs of Staff bitterly opposed such efforts. Eisenhower wanted the JCS to realize "the balance between minimum requirements in the costly implements of war and the health of our economy." He warned that "some day there is going to be a man sitting in my present chair who has not been raised in the military services and who will have little understanding of where slashes in their estimates can be made with little or no damage." "Many other programs" besides military ones "contribute to making the nation strong," he once told Defense Secretary Wilson, "and . . . I myself am probably the only man in a position to bring all these together." In 1956, Eisenhower told John Foster Dulles that he was "getting desperate with the inability of the men there to understand what can be spent on military weapons and what must be spent to wage the peace." But still he let them spend on.[48]

This permissiveness can be explained in part by Eisenhower's belief in many of the basic ideas that characterized the corporate scientists and the wartime managers of state science. Like Bush and Conant, and like Hoover himself, Eisenhower assumed that moderation, balance, and cooperation would somehow in and of themselves prevail over other, more extreme political positions. But that was not proving to be the case with the military and its increasing appetite for spending and proliferation.

Another factor was Eisenhower's belief in the culture of official secrecy. In spite of his enthusiasm for the ideas that Oppenheimer and Bush expressed in their 1952 disarmament panel report, Eisenhower remained a firm believer in the importance of secrecy with regard to nuclear weapons, and Lewis Strauss did nothing to discourage those beliefs. As a result, new perspectives and different ideas never made it over the imposing wall that Strauss erected around the nuclear policymaking process.

In his second term, however, Eisenhower showed signs of disenchantment with the growing militarism in American politics and especially with the scientific advice of Strauss and Teller. In October 1956, he complained that scientists always seemed to be recommending new projects. He wished, he said, that he had some scientific advisers who would recommend things that could be dispensed with.[49] He reminded Strauss in 1956 about his hope "that the need for atomic tests would gradually lift and possibly soon disappear." In more impassioned moments he would blurt out his strongest feelings about the arms race to his aides: "My God, we have to simply figure a way out of this situation. There is just no point in talking about 'winning' a nuclear war."[50]

Strauss tried hard to counter the president's uneasiness. As Eisenhower

began to express qualms about continued American testing and the resulting fallout, Strauss devised a new rationale: he arranged for Lawrence,
Teller, and their Livermore colleague Mark Mills to explain to Eisenhower
that testing was necessary to help scientists devise a "clean weapon" with
little radiation fallout. It was necessary, in other words, to produce more
fallout over the villages of America so as to protect them from more fallout!
But even though he warmed to the idea of a "clean bomb," he also cautioned
Lawrence, Teller, and Mills that he did not want the United States "crucified
on a cross of atoms, so to speak," and that he would continue to seek a
reduction of the nuclear danger.[51]

The Gaither Report of November 1957 was a decisive event in Eisenhower's journey away from the Cold War fanaticism of his advisers. It
provoked the first, faint reconsideration of the proposition that the arms
race could be "won" through continuing technological innovations. Eisenhower convened the panel for the purpose of providing him with recommendations about how to defend against a nuclear war. The committee
members then broadened their brief to encompass a study of the country's
offensive capabilities as well. The panel was formally named the Security
Resources Panel and was led by Rowan Gaither, a San Francisco attorney
who was a friend of Ernest Lawrence's and a nonscientist in the tradition of
Alfred Loomis and Lewis Strauss. But Gaither fell ill and was replaced by
Robert Sprague, a Massachusetts businessman. The panel soon took on
gargantuan proportions: there was a steering committee of eight members,
an advisory panel of nine members including Lawrence, a committee of
members from the President's Scientific Advisory Committee (PSAC), including Rabi and James Killian, and some sixty-seven project members,
including York, Paul Nitze, Spurgeon Keeny, and Richard Bissell.

When the day arrived for him to be briefed about the panel's unanimous
recommendations, Eisenhower sat attentively in the Oval Office ready to
receive the wisdom of the scientists. But he found himself in complete
disagreement with what he heard. The Soviets, the report concluded, had
"probably surpassed" the United States in numbers of intercontinental missiles, and their ICBM capability posed the threat of a surprise attack against
America's SAC squadron. "The USSR will achieve a significant ICBM delivery
capability with megaton warheads by 1959," the report predicted. "The next
two years seem to us critical. If we fail to act at once, the risk, in our opinion,
will be unacceptable." The report, one of whose principal authors was
Strategic Survey and NSC-68 veteran Paul Nitze, recommended a dramatic
increase in defense spending and missile production through 1970.[52]

Eisenhower rejected the panel's suggestions. The report's recommendations regarding increased spending and civil defense construction were so excessive that they led Eisenhower to a conclusion that was entirely outside the assumptions under which the authors were operating. "You can't have this kind of war," he told one of his science advisers, Jerome Wiesner. "There just aren't enough bulldozers to scrape the bodies off the street." The scientists had become such uncritical advocates of the arms race that the president of the United States found himself far ahead of them in his desire for some diminution of the nuclear danger. The advisee had begun to turn the tables and advise his own "experts."[53]

It was not easy politically for Eisenhower to reject the Gaither recommendations. The panel's work was initiated in April 1957, but by the time it presented its report on November 7, the political landscape had been radically transformed. The Soviets had successfully tested the first intercontinental-range missile, which they fired from Tyuaratam across Siberia. The first Sputnik was launched on October 4. A state of hysteria was about to grip the public, and Eisenhower must be commended for his political courage or, at least, stubbornness in refusing to be stampeded by the recommendations of the panel in the heated post-Sputnik atmosphere. Pressure was also being applied by political opponents such as Senate Majority Leader Lyndon Johnson, who called for the release of the report, and by Gaither panel members who leaked their findings to the *Washington Post*, which published a sensationalistic and rather distorted summary of the panel's recommendations.

Sputnik seemed to be a dream come true for the scientific militarists. Here was a prelude to the technological "Pearl Harbor" that the Soviets might carry out; here was the irrefutable argument in favor of increased defense spending, production, and testing. But paradoxically it did not turn out that way. The press and the Congress charged that Sputnik simply proved what the administration's congressional critics such as Senators Stuart Symington, John F. Kennedy, and Henry Jackson had been saying all along: the administration was not paying enough attention to the missile race, and clearly some new blood was needed in the ranks of the presidential scientific advisers.

That is precisely what began to happen. Eisenhower convened a meeting on October 15, 1957, with fourteen of the nation's leading scientists, including I. I. Rabi, who pointed out that the president could benefit from having a permanent scientific adviser in the White House. Eisenhower agreed, and one can only imagine the inner horror experienced by Strauss. His monopoly was beginning to crumble. Two weeks later, on October 29, 1957, Rabi met with the president again with Strauss present. Rabi suggested that a test ban might actually benefit the United States, the Sputnik hysteria to the contrary, because the United States had a technical advantage over the Soviets in regard to H-bomb technology, and a freeze on testing would also freeze that advantage in place. When Rabi left the meeting, Eisenhower kept Strauss and urged him to explore the idea. Strauss, predictably, began to attack it. Among his stated reasons was the "impossibility" of getting Rabi and those aligned with him to work with Lawrence and Teller. "Rabi and some of his group," he said, "are so antagonistic to Drs. Lawrence and Teller that communication between them is practically nil."[54]

This argument echoes the old stereotypes about erratic, egotistical, and unreliable scientists. Strauss knew that Eisenhower liked Lawrence precisely because that famous American scientist was so unlike the stereotype. But Eisenhower had also known and liked Rabi ever since their days together on the Columbia University campus. Rabi, in his own way, strayed from the conventional image. He was down-to-earth, personable, and easy to get along with. Best of all, he never talked down to Eisenhower during the latter's trying tenure at Columbia. Rabi's discussion with Eisenhower gave the president some startling and perhaps even refreshing news: there were actually scientists who disagreed with Lawrence, Teller, and all the other hawkish members of the Livermore faction.[55]

Drawing on a wider range of scientific opinion helped the administration to deflect some of the public criticism it faced as a result of Sputnik. That criticism was something of a shock to Eisenhower, who thought the public shared his levelheaded perception of the Soviet threat; he did not realize how, in James Killian's phrase, "psychologically vulnerable" the public had become. By appointing Killian as his science adviser and by reorganizing the President's Scientific Advisory Committee and making it part of the White House staff, Eisenhower was able to give himself political cover in the face of that public anxiety and anger. It also gave him, Killian noted, the institutional means to begin to pursue in earnest his interests in arms control "against the determined resistance of an Air Force–AEC complex of interests." The PSAC also helped ease the tensions that existed between the administration and the scientific community following the Oppenheimer affair. The membership of the panel reflected the coalitions of the wartime state science community. Edwin Land, a particular favorite of Eisenhower's, was another self-made man and the inventor of the Polaroid camera; Killian was an engineer and administrator in the mold of Vannevar Bush; Kistiakowsky and Rabi represented the academic community.[56]

One-half of the committee's members were administrators rather than research scientists, and many were veterans of the MIT Radiation Laboratory, Los Alamos, and Bell Labs. "Seasoned by their extensive experience in public affairs," Killian later wrote, "these scientists knew how to mix with-out condescension or shyness with high political, military, and industrial personnel. Even though they were intellectuals, they achieved a remarkable impedance match with Eisenhower, the alleged nonintellectual." The panel, Killian further noted, "had the reputation of being more charismatic than bureaucratic." Their primary brief consisted of space and defense matters, but several members "were motivated by what most of us felt to be the higher priorities—the contributions science could make to peace and to the quality of American society."[57]

The PSAC "repeatedly opposed" what York would later call "hard sell" technology. It recommended against the development of nuclear-propelled aircraft, and it attacked some of the Defense Department's command and control systems as "unsound and too expensive." It tried to stop the redundancy of developing both the Thor and Jupiter missiles but failed. It also sought to limit nuclear testing. Killian and the PSAC reinjected some old-fashioned conservative managerial values into state science. The body's members were particularly interested in combating the "near hysteria" in Washington and in the country at large that had taken hold as a result of Sputnik.[58]

The reaction to Sputnik particularly disturbed Killian. It was one of the "strange aberrations of the late 1950s" that suggested how the political culture of the Cold War had gone seriously awry. "The cold war," he wrote, "was already forcing the United States into an arms race and into a position in which the distribution of energies would be decided not by the collective wishes of the American public, individually established by what the ordinary citizen chose to buy and sell, but by cold-war decisions taken in Washington and in the Kremlin, and in a sense imposed upon the individual." He considered it one of his principal responsibilities to nudge the president and other political figures to keep their attention on the political and not simply the technical character of the problems they were considering.[59] Note the consumerist alternative that Killian presents as a norm which the Cold War violated. Consumerism and militarism were still the two sides of the same ideological coin in the minds of many leading scientists and policymakers.[60]

Eisenhower, however, was largely responsible for perpetuating the distorted political climate Killian wrote about. He made it clear in the 1956 election campaign against Adlai Stevenson that nuclear weapons issues were outside the bounds of political discourse. The popular military hero was able to score easily and effectively against his Democratic opponent by belittling or simply ignoring Stevenson's comments on the test-ban issue.[61]

But Eisenhower was not just playing smart politics when he parried Stevenson's thrusts on this matter. He genuinely believed that issues like the test ban were "manifestly not a subject for detailed public discussion for obvious security reasons." Nuclear weapons had become so effectively obscured within the government's culture of secrecy that fundamental questions of war, peace, and public health were deemed beyond the pale of legitimate public debate.[62]

This attitude on Eisenhower's part exemplified the degree to which the scientific militarists and their political allies had been able to define the relation between scientists, political leaders, and the public in a nuclear-armed society. Strauss, Teller, and Lawrence had consistently argued along the ideological lines established by the corporate scientists of the 1920s and by the militarists of the Cold War: basic research in weapons technology was necessary to keep the United States ahead in its evolutionary race against the Soviets, and testing and research were also necessary to find a solution to this race. Secrecy and the absence of dissent were also considered essential to allow this technical process to continue unimpeded.

Certainly no dissent was permitted in the official world of scientific advice that Strauss controlled. But there was growing public protest about the dangers of testing, and Strauss's answers to these protests were losing their persuasiveness to Eisenhower. The AEC chairman actually used a form letter to answer critics and to explain that testing was necessary to achieve a "clean bomb" and an antimissile missile. He also emphasized that a cessation of tests would only weaken the United States, thus emboldening the Soviets and making a nuclear war more likely. "The consequences of any other course," he once argued in support of continued testing, "would imperil our liberty, even our existence."[63]

There was a part of Eisenhower that recognized all of these arguments as lunacy, but he began to act on his sounder impulses only once the most fervent advocates of those positions were either dead or out of the government. Strauss's term as AEC chairman expired in 1958; Lawrence died the same year; and Teller's extreme views on all nuclear questions progressively diminished his credibility. Eisenhower was particularly annoyed when Teller publicly proclaimed that Sputnik was a greater disaster for the United States than Pearl Harbor.[64]

The establishment of the PSAC also seemed to give Eisenhower a much greater sense of confidence in his own moderate impulses. As that confidence grew, Lewis Strauss's days as a scientific adviser were numbered. Eisenhower, especially in his second term, seemed as if he was only waiting for the right push to send him fleeing from Strauss's advice. After an April

1957 meeting with Lawrence, Teller, and Mark Mills, Eisenhower acceded to their recommendations and to Strauss's wishes for continued testing. But after the scientists left he complained to Foster Dulles how Strauss made "it look like a crime to ban tests."[65] Strauss was able to talk Eisenhower out of the president's intention to cancel the "Hardtack" test series in February 1958, but his acquiescence was very reluctantly granted. Strauss justified the need for the tests in the name of his ever elusive "clean bomb," but Eisenhower later told Killian that "he had never been too much impressed or completely convinced by the views expressed by Drs. Teller and Lawrence that we must continue testing of nuclear weapons."[66]

When Eisenhower told Strauss in the spring of 1958 of his intention to announce a unilateral test moratorium, the latter replied that Eisenhower was placing himself in the same camp with Adlai Stevenson and the administration's cast-off disarmament adviser, Harold Stassen. The president exploded, this time while Strauss was still in the room. A commitment to continued testing meant an endless arms race, he said. None of the concerns raised by Strauss—the effect on the labs, the possibility of Russian cheating, the dangers of allowing the Soviets to gain missile superiority—outweighed the importance in Eisenhower's mind of taking this bold step in the direction of peace, rather than continuing along the path of confrontation and possible nuclear war.[67]

Strauss left the meeting realizing that his differences with Eisenhower over this issue might constitute a "permanent and fundamental disagreement" between himself and the president. For Eisenhower, the announcement of a unilateral moratorium, which went into effect at the end of October 1958, was something of a declaration of independence from the mind-set of the Cold War superhawks. He had broken free from the basic assumption of that creed: that scientific research and experimentation must never and could never be checked.[68]

Killian and the PSAC scientists encouraged Eisenhower's latent moderation. The PSAC members also sought to check the nascent fervor to rush into space, where some American leaders hoped to continue the Cold War. Senate Majority Leader Lyndon Johnson, for example, spoke about the importance of obtaining "the ultimate position—the position of total control over earth that lies somewhere out in space." Killian feared this "dangerous innocence."[69] In contrast to such congressional recklessness, the members of Killian's PSAC had a refreshing restraint about them. The sheer location of their offices instilled a conservative sensibility among several of the participants. Since they had been bureaucratically relocated from the Department of Defense to the White House staff, they were given offices in the Old Executive Office Building, a hoary historic structure across the street from the White House. The Departments of State, War, and Navy had been housed there in the seemingly ancient era before Pearl Harbor. In fact, the panel's meeting room had formerly been the anteroom of Cordell Hull's office, where, York noted, FDR's secretary of state "had chewed out the Japanese ambassador immediately after the attack on Pearl Harbor."[70]

Herbert York wrote that "the history of the place and its conservative magnificence made our purpose for being there seem all the more important." In the rare moments when he had time to reflect, he "became deeply aware of having somehow fallen into at least an important eddy in the mainstream of American history." Killian also felt the weight of history in the building's spacious offices, which he contrasted favorably with the "rabbit-warren interiors" that were typical of most government buildings. The purpose of federal office design, he noted, was often to give physical expression to hierarchical distinctions. The Old Executive Office Building offices, however, seemed "to defeat bureaucracy and protocol by surrounding all its occupants . . . with a certain individuality and dignity." This sense of dignity was a matter of political and not just aesthetic importance. Killian and the PSAC represented a refreshing change from the subservient role of scientific expertise devised by Strauss and Teller. But "Ike's scientists" were still far from the vigorous, important public figures that Bush and Jewett in their different ways had hoped scientists might become.[71]

Killian tried to continue the intellectual approach to state science advanced by Bush: scientists were to be partners but not special advocates. In a 1959 speech to the American Association for the Advancement of Science he said, "The current emphasis on science, if it is not to cause reactions adverse to science, also requires of the scientific community humility and a sense of proportion. It requires of scientists a recognition that science is but one of the great disciplines vital to our society and worthy of first-rate minds, a recognition that science is a partner—sharing and shouldering equally the responsibilities which vest in the great array of professions which provide the intellectual and cultural strength of our society."[72]

By 1960, the scientific elite was unwilling and unable to forge a new direction for American political culture, as earlier elites had tried to do. "For good or ill," Killian wrote, "I had no instinct for power and preferred to be 'a private man in the public society.'" He was not bent on forging a full-time career for himself as a presidential adviser. Unlike Strauss, he did not lust for power and continued access. Therein lay Killian's great moral strength but also his greatest political and bureaucratic weakness. He saw the same traits in his fellow PSAC members. "A majority of its members," he wrote, "had no career objectives in government. . . . PSAC members always sought to be non-partisan, whatever their private political beliefs might have been. They would have rejected as repugnant and ridiculous the idea that they were a 'priesthood,' a term that has been used to describe the scientists who undertook public service in that period. They did not consider themselves part of an 'establishment.' "[73] But if they were not part of the "establishment," were they independent experts? Could they be effective critics of the government's military and scientific policies? By the end of the 1950s, the professional culture of scientific militarism and the political climate of the Cold War state made it very difficult for state scientists to speak out on issues such as proliferation and testing.

The career of Herbert York is illustrative of this problem. He would become a prominent critic of the arms race, but his vision of the role that scientists could play in American politics and culture was much narrower than that of earlier state scientists such as Bush and Conant. It was tremendously difficult for York to envision *any* role for himself as a public critic of government policy since his entire professional identity had been formed in the institutions of state science that had taken shape during World War II. He was part of the first generation of scientists who had no prominent professional affiliations outside of state science itself, and his connections to that world came through Ernest Lawrence, who installed York as the first director of the Livermore Laboratory. York acknowledged that he owed everything that he was professionally to Lawrence, and that connection and its dissolution are crucial to understanding his behavior as a public figure.

York arrived at Lawrence's Berkeley laboratory in 1943, fresh from his undergraduate studies at the University of Rochester. York, like so many of the other leading lights in state science, came from a relatively humble background and the ways in which state science whisked him up the social scale are important to consider when trying to understand his loyalty to that system. His father was a railway express messenger who fortunately had a love of learning and instilled a healthy intellectual ambition in his son. "From the earliest times," York wrote, "I remember his saying he did not want his son to be a railroad man. He made it clear that that meant I should go to college, even though he knew little about what that actually entailed."[74]

York found the world of higher learning incredibly exciting. "I still remember vividly," he wrote of his initiation into university life at Rochester in 1939, "the discovery of all manner of things I had previously been totally unaware of. The joy of learning how the world actually worked, the existence of such things as graduate students, Ph.D. degrees, and, above all, mature people who did what they were doing because they enjoyed it and not just to earn money to support their families. College, I found, meant a great deal more than simply not working on the railroad, and I was determined to become a part of this newly discovered world."[75]

Lawrence was York's guide into the exciting new world he had entered, and the young student quickly and willingly adopted the prejudices and viewpoints of his mentor. He learned, for example, that Lawrence eschewed political activism among his students. They were to leave the higher authorities to deal with the larger social and political implications of their work. "While he also believed that scientists should give advice when asked," York wrote in his memoir, "he openly discouraged voluntary staff involvement in such broader issues. Scientists, he said, especially young ones, should not waste precious working time on extraneous issues for which they had no special training. . . . Discussions about nuclear political issues, which I later learned were commonplace at some other laboratories, were frankly discouraged at the Rad Lab."[76]

Lawrence, according to one of his biographers, "largely dismissed perplexing events in the troubled world outside the laboratory from his mind; they would be worked out by those responsible, at least in America." He once told Robert Oppenheimer, "You're too good a physicist to get mixed up in politics and causes." All of his students knew precisely where their master stood on questions of science and politics. "If anyone wants to write letters to the editor and that sort of thing," Lawrence told them, "he should get out of science and get a job on a paper." The prestige of a politically engaged scientist did nothing to impress Lawrence with the wisdom of such engagement: for example, he scorned Niels Bohr's passionate efforts to devise new political controls for nuclear weapons.[77]

He seemed to think that devotion to science, in and of itself, removed any motive for seeking out the fellowship and commitment offered by political affiliations. In a revealing statement to Frank Oppenheimer, Lawrence sought to argue the latter out of his leftist commitments by saying, "Good scientists aren't like people who just want to eat, sleep, and make love. You're not like people who can't get anywhere. You don't need that!" For Lawrence, politics was the last refuge of a loser. Professionalism and personal success, by his way of thinking, simply abolished the need for politics.[78]

Aside from his philosophical objections to politics per se, Lawrence also feared the potentially negative consequences that could come from having his beloved lab embroiled in conflicts. In 1948, when Frank Oppenheimer was quoted in a newspaper explaining that a political speech he had given was held in a small auditorium because the owners of a larger hall would not permit an integrated audience, Lawrence expostulated, "Now look what you've done! You've brought race relations into the lab!" Lawrence had helped Frank get a job at the University of Minnesota after the latter had been railroaded out of the Rad Lab, and Lawrence seemed to interpret Oppenheimer's continued leftist commitments as some sort of "betrayal." Relations between the two disintegrated, further straining the relations between Lawrence and Frank's brother, Robert Oppenheimer.[79]

The only politics that mattered to Lawrence was the internal politics of the lab, where his power and authority were uncontested. York felt the consequences of slighting this authority on a few occasions: he once transferred himself from a duller to a more exciting project without getting Lawrence's permission. "He glared at me fiercely," York recalls, "and, with his jowls quivering, said, 'You had better learn which side your bread is buttered on if you want to remain in this laboratory.' " In another incident, Lawrence had baldly told York to cooperate or get out. Others, such as the noted Italian physicist Emilio Segre, told York that these were merely the temper tantrums of a great man and were not to be taken too seriously. York went along with Lawrence's strictures and conformed to his demands. He accepted the notion that a scientist should keep out of politics and leave such matters to the "responsible" authorities.[80]

Aside from his directorship of Livermore, York participated in several advisory panels in the late 1940s and 1950s. He was a member of the air force's Science Advisory Board, the SAB Nuclear Panel, the Strategic Missile Evaluation Committee, the Scientific Advisory Committee on Ballistic Missiles, and the Gaither panel. He was greatly impressed with the brilliance and charm of John Von Neumann, almost as much as he had been with Lawrence. The advisory panels that Von Neumann led were always characterized by the spirit called for by the Gray Board: a single-minded dedication to the maximization of the offensive capability of the United States. York uncritically accepted the notion of a permanent superpower rivalry and arms race that guided the work of these bodies. He believed he was serving his government best by helping it wage that race more effectively.

He remained silent during the Oppenheimer affair out of devotion to Lawrence and a desire to protect the new Livermore Laboratory. "I was extremely fond of [Lawrence] and felt I owed everything I was professionally to him." He simply could not go against his mentor in this matter. Besides, a break with the vindictive duo of Teller and Strauss "would have very seriously imperiled the whole [Livermore] endeavor. . . . I still think that I did right, considering what the circumstances were and how I comprehended them, although sometimes on rainy nights I have some doubts."[81]

York could not help observing that the stress of trying to fend off the political controversies of the day served to exacerbate Lawrence's health problems. Lawrence underwent a tragic deterioration in a relatively short period of time from 1952, when he was diagnosed with ulcerative colitis, until his death from that condition in 1958. Medically, at least, his decline could easily have been slowed and even reversed had the patient showed the slightest inclination to cooperate with his doctors or to heed the pleas of his wife and close friends. Lawrence had always been a tireless worker who drove himself mercilessly. In his youth that energy was the essence of his famed "boyish exuberance" and enthusiasm. But his insistence on maintaining a frenetic schedule of work and social commitments in the 1950s in the face of overwhelming medical evidence that he needed to rest above all else casts these last years of his life in a disturbing hue. He was unable or unwilling to discontinue habits and characteristics that no longer seemed boyish and wonderful but were instead dark and almost suicidal.

His increased drinking in these years exacerbated his problems. Even his worshipful official biographer, who insists that Lawrence did not drink excessively—for a man of his size—could not help piling up the examples of his alcoholic tendencies after 1950. He refused to give up drinking at any point during his last six years, when doctors repeatedly told him that alcohol was exacerbating his intestinal troubles.[82] Lawrence's temperament deteriorated along with his health. The outbursts of anger became more frequent and uglier. His decline reached its nadir when, à la Patton, he slapped a subordinate and had to make a remorseful public apology to the entire laboratory staff a few days later.[83]

York's membership on the PSAC during Lawrence's last years exposed him to new political and professional perspectives and to a world that was wider than Lawrence's Radiation Laboratory and its satellite, Livermore. York had gained entree into the PSAC as a result of the official positions he held in the late 1950s: head of the Advanced Research Projects Agency (ARPA) and then as assistant secretary of defense for research and development. The members of the PSAC were "men whose views about the world in general and national security in particular were considerably broader and more complex than those of the people I had previously been close to." But even Lawrence himself, near the end of his life, seemed to be moving away from the hard-line, Cold War policies he had been advocating throughout the 1940s and 1950s. During his last visit with his mentor, York was surprised to hear him express skepticism about the increasing power, size, and cost of the accelerators that were being constructed around the world. There was no paternal pride in Lawrence's discussion of this subject. Rather, he seemed to be wondering out loud to York if the work of the cyclotroneers had outstripped its economic justification.[84]

His work on the early accelerators in the 1930s, Lawrence explained, was carried out with the clear understanding of the connection between research in particle physics and the effort to split the nucleus and unleash tremendous power. There was a clear practical application to the work. Yet in the postfission and postfusion world, the accelerators did not seem to have any practical applications. Certainly they could continue to be excellent research instruments for particle physicists, but that did not seem to justify the commitment of huge amounts of public or private money to support such esoteric research.

York found it "quite surprising" to hear the "father" of big science subject his own field to such social and political considerations. "Lawrence," York realized, "was definitely moving away from his previous totally uncritical view of accelerators; maybe he was doing the same with nuclear weapons." Tragically, however, Lawrence did not live long enough to continue on that intellectual journey toward new ideas and new perspectives. The damage he had done to himself was beyond repair. While serving in Geneva at a disarmament conference he experienced a devastating colitis attack. He was also simply exhausted. His worsening condition forced him to fly back to the United States for emergency surgery that would have removed his colon. He died on the operating table, only fifty-seven years old.[85]

The other remarkable man who influenced York's participation in state science, John Von Neumann, also met a premature death, in his case from cancer. York's memory of Von Neumann's last days also suggest that other, more human considerations were crumbling away the thick ideological wall behind which York had uncritically labored in support of the arms race. He visited Von Neumann in the hospital, accompanied by Lewis Strauss, and tried to divert him by starting a conversation "about some technical topic." Von Neumann, who had become "despairing" and had turned to Roman Catholicism for solace, "would say no more than a simple hello before turning to Strauss to discuss some detail of his will."[86]

After the deaths of Lawrence and Von Neumann, York's thinking began to move in the direction he had perceived during the final conversation with his mentor. It was a slow process, and there were many ideas in York's mind that militated against the more critical and even radical implications of the "new thinking" to which he became exposed after Lawrence's death.

Even when York did become a full-fledged dissenter, he remained a loyal supporter of the culture of the insiders. Like Jewett and Bush before him, York was intellectually and ideologically hemmed in by the political, institutional, and cultural values that he had imbibed in the course of his work.

He took an important step away from an uncritical Cold War mentality at an April 1958 conference at Ramsey Air Force Base in Puerto Rico. York and other PSAC members were considering the question of whether a nuclear test ban was in the United States' interest and whether it could be effectively monitored. He abstained from voting on either issue. "I argued that the matter before us was essentially a political and strategic issue and that a group made up entirely of scientists wasn't appropriate for deciding such questions. There must be, I felt, some other group somewhere else in the government better suited for dealing with such matters." Jerome Wiesner took York aside and "patiently explained several things" to him, namely, that the president was entitled to ask anyone he wanted for advice and that "there really was no one else; it was us or no one, be that plausible or not."[87]

By 1960, York realized that the ceaseless escalation of the arms race did not bring greater security. Paradoxically, it was the quest for "security" itself that constituted the chief source of danger and insecurity in the world. He grew increasingly frustrated with the Cold War fanaticism that poisoned state science in the Eisenhower years, and this new attitude reached a personal culmination in 1960, when he found himself being called a traitor by John McCone for daring to express the view that there was no intelligence to support the notion that the Soviets were violating the informal nuclear testing moratorium that was then in place.[88]

Two conversations in 1960 also had an important role in shaping the new direction of York's thought. The first was with General Laurence S. Kuter at the North American Air Defense Command in Colorado Springs. The second was a briefing at the Strategic Air Command headquarters in Omaha, Nebraska. Kuter and York were discussing the impending implementation of the Ballistic Missile Early Warning System (BMEWS), which would, they hoped, provide fifteen minutes' notice of an incoming nuclear attack. Kuter mentioned the need for achieving a high level of reliability in the warning system because BMEWS was going to be connected to the launching mechanisms for the American ICBMS.[89]

York was stunned. This represented a "launch on warning" strategy that not only eliminated the president from the decision-making process but eliminated human beings entirely. Nuclear war would be automated and could conceivably occur as a result of technical error. When York told Kuter that under no circumstances would the launch and warnings systems be linked, the general huffed, "In that case, we might as well surrender now." York replied that surrender was "not the only, and certainly not the proper alternative."[90] Such alarming discussions were leading York to the belief that "the steady advance of arms technology may be leading us not to the ultimate weapon but rather to the ultimate absurdity: a completely automatic system for deciding whether or not doomsday has arrived."[91]

At a SAC briefing in Omaha concerning the likely damage that would be inflicted by an American nuclear attack against the Soviet Union, one of the transparencies showed a lethal level of radiation hovering over noncommunist Finland. General Thomas Power acknowledged the problem but hastened to point out that "this is the dose that some damn fool who went outdoors and looked up at the sky for two days would receive." The cruel sophistry of his analogy offended and disturbed York. He noted how his briefers seemed to assume that "it was the victims in Helsinki who were the 'damn fools' in this situation." Power's attempt to shrug away the moral consequences of nuclear fallout "struck deep." Power's and Kuter's comments "unintentionally made it crystal clear there is something seriously wrong, and at bottom absurd, about maintaining peace through mutually assured destruction."[92]

By the time he was preparing to hand over his official responsibilities to John McCloy in 1961, York had sharpened his thinking on the arms race. He had reached three important conclusions: that defense of civilian populations was impossible in a nuclear-armed world, that there was no technical solution to the dilemma of maintaining a nuclear arsenal while averting a nuclear holocaust, and that "our only real hope for the long run lies in working out a political solution."[93]

One year after drafting those points for McCloy, York reiterated them in testimony before the Foreign Relations Committee in 1961: "Ever since shortly after World War II, the military power of the United States has been steadily increasing; over the same period the national security of the United States has been rapidly and inexorably diminishing. . . . It is my view that the problem posed to both sides by this dilemma of steadily decreasing national security has no technical solution. If we continue to look for solutions in the area of military science and technology only, the result will be a steady and inexorable worsening of this situation."[94]

It was becoming clearer to York that scientists would have to act politically to impede the arms race. He realized he could not simply wait for the "responsible authorities" to take things in hand. This insight was underscored for him during a relatively trivial episode with Eisenhower, concerning two air force generals in Florida who were making public predictions about rocket launches that Eisenhower feared might prove embarrassingly overstated. The meeting with the president took place around the Cabinet Room table, where the secretary of defense, the chief of staff of the air force, and his deputy were also present. Eisenhower pounded the table: "Why don't they make those two generals shut up?" York was "amazed by the president's remark. Here *they* were, sitting around this very table, the superiors of those two generals five layers deep and extending all the way up to the commander in chief himself, and yet this they wanted some other they to take care of the problem."[95]

It was one of those "special moments" that left York with a vivid sense of how things really worked in Washington. There was, he believed, no special place at the pinnacle of American power from within which the "big picture" could be seen. This obviously had disturbing implications with regard to the formulation of nuclear policy. In the confusion of political discourse and decision making, were all views, and the most intelligent views, being heard? It was the beginning of a realization in York's mind that there was a vacuum of expertise and leadership into which the scientists would have to move.

But York's political ideas and values were even more nebulous than had been Bush's, Conant's, or Jewett's. He seemed to lack any coherent vision of the role of science or scientists in American life. He knew what scientists should not be doing, but he seemed to have no ideas about what positive, assertive actions they could take in the public sphere. He had been admitted into the rarefied echelons of state science at such a young age that he had never known any institutional life outside it. When York began writing impassioned jeremiads against the arms race in the late 1960s and into the 1970s, he certainly sounded more radical and more critical than Bush ever did. Yet, oddly, his social and political vision was narrower. He had a more comprehensive knowledge about the excesses inside the Pentagon and the dangers these posed to America and the world, but he lacked any language or ideas about what might supplant or replace the culture of the insiders.

Even in his most notable public stance against the arms race, an October 1964 article in *Scientific American* coauthored with Jerome Wiesner, York was highly circumspect in proposing solutions. He was content with arguing that the premises on which the existing system operated were false. He and Wiesner were admittedly concerned with a specific issue in that piece, the test ban. But in all of York's subsequent writings, he gave little attention to articulating a broad vision of how scientific expertise in America could affect the nation's political and social problems.[96]

The connection between scientific experts and the public seemed finally to have broken completely after decades of stretching and strain. Scientific elites had retreated further and further from any direct engagement with the public since the seemingly archaic days of Henry Towne and his quaint ideas of engaging the public intelligence in a museum. Finding professional homes in the corporations and then in the maze of the Pentagon, scientific leaders like York and Killian had lost the ability to speak, in their official capacities, to anyone other than their professional or institutional colleagues.

To be sure, all the scientific leaders I have examined in this book viewed the public as a subordinate player in their political visions. But certainly there was no more passive role for the public than that envisioned by the advocates—and even the critics—of scientific militarism: the public was nothing more than a potential casualty. York recoiled from that ideology when his mind's eye began to see the human beings that were to be, in Eisenhower's grim image, scraped off the streets of a nuclear-blasted world. But how to change this? He could not probe deeper toward a truly new configuration of the relationship between technical experts, the state, and the public.

The proper role of an expert in a democratic, representative government remained a problem for him. "There is a real paradox here," he wrote in his 1987 memoir. Physicists had "special knowledge," which, fortunately, prompted many of them to take a prominent public role in warning the public about the dangers of the arms race. "Physicists and physicians," he wrote, "understand better than others the thermonuclear horror that is always only thirty minutes away from happening." The warnings of these professionals were "correct and soundly based," but "their prescriptions of what to do about the problem . . . are often naive and based on false notions of how things are and why they are that way. Scientists and engineers do not in fact understand the political and strategic elements of this issue as well as do political scientists, Sovietologists, historians, strategists, statesmen and certain others." Unfortunately, those experts often did not fully grasp the horrors of the nuclear war scenarios they could discuss so abstractly.[97]

York believed there were people who passionately wanted to change things but who were naive and neglectful of the real-world difficulties in achieving that change; and there were people who understood the realities and yet lacked the passion or the will to change them. Here is another variation of the dilemma between experts and the public, between specialized knowledge and social change, that was part of American political culture throughout the twentieth century. York saw virtually no role for the public in penetrating the discourse of the experts and the insiders.

The lack of a vocal, critical voice in the government on scientific and military matters was especially injurious during the so-called missile gap controversy. The efforts by Killian, York, and the PSAC to moderate the arms race and to restrain American missile proliferation were completely overwhelmed by this bogus problem. But Democrats in Congress played the red peril card and called for an escalation in the size of the American arsenal even though the United States had an overwhelming nuclear advantage over the USSR.

Eisenhower, who knew the truth about America's clear strategic superiority, refused to dignify his critics by engaging them in public debate or even by peremptorily refuting them. He rejected the idea of public discussion on matters pertaining to nuclear strategy. He remained silent in the face of political attacks, trying perhaps to project an air of dignified confidence and command but inadvertently reinforcing the notion that the old soldier was no longer up to the job. These insinuations clearly got under his skin. "I am always a little bit amazed about this business of catching up," he once said in a press conference. "What you want is enough, a thing that is adequate. A deterrent has no added power, once it has become completely adequate, for compelling the respect of your deterrent." Eisenhower's comment showed the extent to which he was still in agreement with Oppenheimer. The twenty thousandth bomb in the U.S. arsenal would not offset the Soviets' two thousandth, Oppenheimer had noted. But this simple truth of nuclear strategy sailed over the heads of a crude, simplistic press and a political opposition that could not see past the primitive categories of a "race" and a "competition" and could not let go of the unalienable American truth that having more and bigger things was always better.[98]

Eisenhower grumbled behind closed doors, but he never made his overwhelmingly convincing case to the public. He refused to participate in what he thought was a cynical political campaign. In an early 1959 meeting with the PSAC, Eisenhower commented "on the way irresponsible officials and demagogues are leaking security information and presenting a misleading picture of our security situation to our people. Some of our senators in particular seem to be doing this." He referred, specifically, to Senator Symington's close relationship with the vice-president of Convair. Symington's information about the supposed gap was coming from low-level people in the Pentagon and the CIA who did not have access to the U-2 photographs that refuted the existence of any gap in the Soviets' favor. Eisenhower complained to Republican leaders in 1959 that he was "getting awfully sick of the lobbies by the munitions" companies. "You begin to see this thing isn't wholly the defense of the country, but only more money for some who are already fat cats."[99]

Scholars such as Robert Divine and McGeorge Bundy have properly taken Eisenhower to task for failing to give the American public a full, forthright exposition of the nuclear situation. The president chose not do so as a result of the poisonous atmosphere of secrecy that had infected too much of state science and American political culture generally. He feared that a speech that in any way proved America's nuclear superiority over the Soviets would compromise the intelligence techniques which had garnered that proof, especially the u-2 overflights of Soviet territory and the even more secret new spy satellites that were soon to replace the u-2.[100]

Eisenhower fended off the urgent pleas of Christian Herter, Foster Dulles's successor, and presidential candidate Richard Nixon to release the evidence gathered by the u-2s. He did, however, agree to the remarkable arrangement of letting Jerome Wiesner work as an adviser to the Kennedy campaign in 1960 while Wiesner kept his position on Eisenhower's PSAC. Wiesner told the historian Gregg Herken that in spite of a presidential admonition against divulging classified material, he believed Eisenhower hoped that he would discreetly disabuse the Democratic candidate of his faith in the missile gap's existence.[101]

The perpetuation of the missile gap controversy was one of the great tragedies of the Eisenhower administration and of the entire Cold War era. It arose just at the moment when Eisenhower was beginning to act on his long-held reservations about the arms race, just at the moment when he was bringing back into the White House men like Killian, Wiesner, and York, who could give him sound advice about scientific matters. But in a political climate poisoned by the hysteria over Sputnik, the Gaither leak, and, more fundamentally, the persistent, insidious belief that America's security lay in relentlessly keeping ahead of the Soviets, a crude accusation like the "missile gap" inevitably seized the public mind and dominated the 1960 electoral campaign.

At last, in 1961, Wiesner was free to brief the newly sworn-in Democratic president about the incontrovertible evidence against the existence of a Soviet advantage. Kennedy's one-word response, spoken not in relief but in anger, was an expletive. He never publicly acknowledged the error of his campaign attacks.[102]

The 1960s were a paradoxical period for American scientists. The ideas of the corporate scientists and state scientists were crucial aspects of America's governing ideology, yet scientists themselves were not central figures in the state. Big science had joined with big business and big government over the first half of the twentieth century, but the cost of those alliances was the steady suppression of scientists' voices as public figures and the redirection of their political ambitions toward the interests of their institutional patrons.

Men like Henry Towne, Frank Jewett, and Vannevar Bush wanted to use the cultural, intellectual, and economic power of science to achieve particular political goals. For Towne, it was a stable, conflict-free class system. For Jewett, it was a social order in which scientists were able wisely and responsibly to shape social change without the ham-fisted interference of government. Bush, along with colleagues like James Conant and David Lilienthal, wanted to establish a partnership between science and the government—and, more broadly, between professionals in the private and public spheres—to create a society that was prosperous, militarily strong, and meritocratic.

But when scientists entered the corporations and the government, they had to accommodate their ambitions to the agendas of their institutional patrons. AT&T executives did not share Jewett's conception of the public role of scientists. Generals and political leaders in the federal government did not share Bush's view of the scientists' role in shaping policy. In the corporate culture, scientists were simply to produce the marvels that brought prosperity and (presumably) social harmony. In a militaristic, Cold War culture, scientists were to be mere "technicians," providing the military with the weapons it wanted.

Paradoxically, scientists like Bush, Conant, and Lilienthal created the very ideas that were used to weaken their position in the government and the policymaking process. The social vision of these men stressed a supposedly ineluctable process of technical evolution that had to be managed by collaborating elites for the political, economic, and military benefit of the American public. To them, the sheer act of cooperation and the pooling of professional and institutional resources would create a smoothly working process of progressive change, and their brand of progress was defined as more technological innovations that expanded prosperity, created more opportunity, and enhanced the nation's military strength. These ideas had powerful resonance among the generals and politicians who embraced the technical and social promises of scientific innovation. Vannevar Bush's correspondence with Omar Bradley was illustrative. Bradley was all for using science to get bigger and better weapons in the American arsenal. He was not particularly interested in Bush's insistence that scientists had to participate in the decisions concerning the use of those weapons.

People like Strauss and Teller eagerly took the ideas of Bush and even those of the corporate scientists of the 1920s and gave dangerous new meanings to concepts like technical evolution, elite collaboration, and public good. In the face of the Soviet challenge, Strauss believed, technical innovations were great risks that had to be pursued before the Russians also pursued them. Elite collaboration was no longer a matter of partnership but merely subservient loyalty to military imperatives. And the public good was no longer defined in terms of abundance and social harmony but solely in the grim language of military security and survival.

The adaptation of the ideas of corporate science and state science toward the imperatives of Cold War militarism was made easier, however, by the crucial political choices of the World War II scientific leaders: they chose to be deferential, noncombative partners to the military and political elites. They participated in, and helped strengthen, the culture of official secrecy that proved so poisonous in American science and politics as the 1950s wore on. But most important, they saw themselves as public figures who had little engagement with the public. We saw this in the writings of Bush, Conant, and Lilienthal. The public was to be a mere passive spectator and grateful recipient of the scientific and institutional innovations provided by the experts. When the institutions and ideas of state science were turned against dissenting scientists like Oppenheimer during the 1950s, Bush, Conant, Lilienthal, and Arthur Compton were furious, but they were incapable of taking any effective public action against this process. They had made themselves "insiders" who did not make a "ruckus." They were "good soldiers" and could not reach outside their insider culture to set forth ideas or principles that would have challenged the outrageous, at times illegal, actions of men like Strauss, Lawrence, and Teller. When Herbert York became alarmed and disillusioned over the direction of state science and even when he became a vocal critic of the arms race, he could not formulate a clear idea of the public's role in addressing the problem. The public was still a passive spectator. The difference between Teller and York was that Teller saw the public as passively protected by America's expanding nuclear arsenal; to York the public was passive but endangered by the threat of nuclear oblivion.

York's career shows the extent to which scientists in government were insulated from the public. But when York is compared with Henry Towne, an even greater historical change is evident. York not only had a radically different conception of the public role of scientists and engineers, his entire social and political reality was different from Towne's. Where Towne saw his role as that of a pedagogue, instructing the public about the benefits of science and the proper relation between classes in a technological society, York was a creature of the state who used his expertise in the service of the military. Towne saw himself as a prominent and independent figure in a particular community. He was not affiliated with a corporation, a government agency, or a national network of professionals and experts. He wanted to use his expertise to achieve political harmony between social classes. York (when he was director of Livermore) was uninterested in such sociological concerns and devoted himself instead to pressing at the technical boundaries of his discipline.

Even when York became a public figure, his manner of addressing controversies was very different from Towne's. The latter always worked on the assumption that common understandings had to be achieved between experts and the public, between employers and workers, leaders and led. That was what Towne was trying to achieve in his museum and in various proposals for reducing class conflict and political tension. But by 1960, a leading scientist such as Herbert York did not address the public merely on the basis of his intellectual authority as a scientist. He could act only as part of the professional and institutional communities of which he was a member. Furthermore, York himself could only envision social change coming from within that professional and institutional complex. Even when addressing the public dangers of potential nuclear war, he could not envision the public as a possible force for change. Rather, he came to the conclusion that he and other elite insiders had to act within the government against this danger.

So York was left in the 1960s at the same place where Bush had been left in the 1950s: pondering the paradox of the insider. Both men believed their greatest effectiveness lay in collaborating with political and military elites, so how could they criticize those elites while remaining trusted and effective insiders? There is tragedy as well as great irony here, for scientists like Bush, Conant, and Lilienthal wanted to use their partnership with the military and the state to achieve some great and laudable goals. By enhancing the social and political power of scientists, they wanted to spread American prosperity, and they wanted to expand the opportunities for professional advancement beyond the narrow fraction of Protestant, northeastern elites that had enjoyed them before World War II. In many ways they succeeded. America did become more prosperous. Educational capacities expanded dramatically as a result of military funding of colleges and universities, but at the cost of further strengthening militaristic imperatives in American political culture.

Their successes brought more ironic, unintended consequences. The public that benefited from greater prosperity and opportunity in the late 1950s and throughout the following decades began to ask questions about the rectitude and the powers of the professions, the corporations, the universities, and the government itself. The orientation of the American state toward militarism and corporate expansion came under brief and only partially effective attack in the 1950s and 1960s, but the attack was enough to shatter the public's deference toward scientists and experts in general. The problems of the 1950s and 1960s also revealed a crucial weakness in the ideology of corporate science and state science: it was not capable of providing ideas and solutions to fundamental problems of justice and power.

We saw how the emergence of corporate science and state science in the 1930s and 1940s redefined the problems of American politics as simply matters of how best to increase corporate production, how to provide more innovation, more abundance, all the while letting the corporations, the scientists, and the experts preside over this process. Questions about inequality, poverty, and the unequal distribution of power in the society were simply swept aside by the rhetoric of progressive, evolutionary innovation that flowed from the pens of corporate publicists in the 1930s and scientific administrators such as David Lilienthal and Vannevar Bush in the 1940s.

But Americans in the 1950s and 1960s were now addressing problems that were outside the intellectual boundaries of corporate science and state science: What to do about the persistence of poverty and racial discrimination in an abundant, affluent society? What to do about the dangers of militarism in a nuclear-armed world? The ideas of corporate science and state science were inadequate for handling those issues. Consider the writings of James Webb, who was President Kennedy's appointee to head the National Aeronautics and Space Administration (NASA). His ideas represent a continuation of the key themes of corporate science and state science, and he saw his role as a scientific administrator in much the same way that Lilienthal and Bush saw their roles. Indeed, Webb's 1969 book *Space Age Management* reads like a catalog of all the ideas of corporate science and state science that had accumulated since the turn of the century. But by the 1960s, those ideas seemed utterly impotent and irrelevant to the problems facing America.

Webb stressed one of the central ideas to emerge out of the ideology of corporate science: Croly's notion that individual and collective goals could be fused within the new organizational structures of American life. Webb conflated this idea beyond the corporation and even beyond the space program and saw it as the essence of *all* of American history. "The basic concept of our cooperative constitutional system," he wrote, "was that individually free men would become collectively responsible men." He echoed Lilienthal's TVA-inspired notions about management and democracy when he wrote, "What we have learned about these air and space systems, their common elements, and the very different ways they make use of the forces of nature seems to provide an analogy to the new kinds of management action and control systems we need to solve problems that we now face in urban life or to achieve economies of scale in programs for undeveloped regions."[1]

Like Bush, he emphasized the importance of professional and institutional collaborations and patted himself on the back for NASA's success in forging "a working partnership between universities, industry, and government." Webb could even be palatable to the scientific militarists since he also subscribed to their view that America was caught in the grip of a "technological revolution that is now so fully upon us [and which] involves all areas and disciplines. It is the most decisive event of our times, and keeping ahead of it is essential." Most important was the awareness that Americans were in the midst of a "crucial and total technological contest with the Soviet Union."[2]

Sometimes his work reads like a satire on technocratic thought, A Modest Proposal for the Efficient Management of Democracy. He could use phrases like "legislative–executive–public sentiment complex," and he was quite straight-faced when he wrote, "One of the severest trials of those responsible for the [large-scale] endeavor is to maneuver through the turbulence generated by the phenomenon of dissent, which is a ubiquitous part of our political environment." He further intensified the culture of scientific militarism by justifying NASA in terms of its military usefulness—something that Eisenhower and even Kennedy's own defense secretary, Robert McNamara, had refused to do. He urged Kennedy, a few weeks before the president's death, to make sure that the military benefits of NASA were stressed during the upcoming 1964 presidential campaign.[3]

And yet another familiar theme runs through Webb's writings and his career at NASA: the use of state science to open up wider avenues of opportunity in American life. In this regard, Webb was the most passionate of any of the scientific administrators considered in this book. He was, as Walter McDougall noted, an out-and-out social planner.[4] Where Bush used OSRD money to make the elite institutions richer, Webb used the contracting powers of NASA to strengthen the schools of the South and the Midwest. The NASA fellowships were also (in his words) "designed to break the lock that the Ivy League had on the very bright graduate students in the country." He wanted to distribute money so that students could have "freedom of choice" about where they studied; they could go to Oklahoma, Arkansas, or Louisiana, he said, instead of just Harvard, MIT, or Cal Tech, and not "feel like second class citizens." Webb, like Bush and Conant, conceived of his institutional reforms on a national level. The regional colleges had to be improved not just for their own sake but also "to get somehow a proper connection with the profit-making stream of American business."[5]

The space program was no more successful in fulfilling its hopeful vision of American life than had been the atomic power enterprise or the Tennessee Valley Authority. It was as much a nostalgic exercise as it was a forward leap into a bold new world. It was an attempt to reclaim a fading sense of the unifying power of what David Nye called the "technological sublime." That sentiment had exerted a unifying cultural force with the introduction of such earlier technologies as electricity, the railroad, and even bridges. But in a country that was bitterly divided, such easy unity was no longer possible. It was also no longer possible to assert the unambiguous goodness of American science and technology in a time when those forces were wreaking havoc in Vietnam and on the quality of life in the United States. The cold, barren surface of the moon was the fitting destination for the ideological and institutional momentum that had been driving state science throughout the 1950s and 1960s.[6]

The elite scientists and scientific administrators could not think their way toward a novel formulation of the role of expertise in American political culture. They believed that their political and social goals could be achieved only by working through large corporate or governmental institutions and technical projects that were, over the course of the twentieth century, increasingly large, complex, politically embattled, and removed from the day-to-day experience of Americans. Technologies such as the telephone and the light bulb had a clear and direct impact on the lives of Americans, and it was when these technologies were taking hold, in the 1920s and 1930s, that American scientists enjoyed their greatest popularity. Nuclear power, rocketry, and space flight, in contrast, constituted powerful spectacles that seemed to affect ordinary life only in a dangerous and threatening manner, as evidenced by nuclear meltdowns, aeronautic disasters, and the danger of nuclear war.

By the 1960s and 1970s, the American public was no longer convinced of the benign, progressive nature of scientists or their creations. And they were certainly not as deferential toward the national institutions in which many scientists worked. And when the war in Vietnam and race relations became central issues in American politics, the now aged scientific leaders of the 1940s and 1950s aligned themselves on the side of militarism and reaction. Bush, Conant, and Lilienthal supported the Vietnam War in word and deed. Bush and Conant joined other creaky Cold Warriors in the Committee on Peace with Freedom in Vietnam, and Lilienthal was drawn into the war in an even more spectacular way.

Lyndon Johnson, desperately looking to justify the war to himself and to the public, enlisted Lilienthal to visit Vietnam and formulate a TVA-style plan for the Mekong River Delta region. Nothing better illustrates the ways in which the ideological strands of science and militarism remained fused in the minds of the nation's political elites than this escapade. Lilienthal complied, of course, having perhaps inured himself against the use of his talents on behalf of iniquitous causes through his long and profitable business relationship with the Shah of Iran, which lasted from 1955 to 1978.[7]

He made four trips to Vietnam between February 1967 and April 1969, received the usual Potemkin tours, and came back to report glowingly what Lyndon Johnson wanted to hear: there were "fantastically productive resources" in the delta, and it was "like Texas with a lot of water and no oil. And there may be oil too, for all we know." Johnson also told Lilienthal before one of his press briefings: "Dave, you give them some of that philosophy, that good TVA philosophy. . . . As much as you want."[8] Fantasies of a Mekong Delta Authority petered out against the realities of war and Lyndon Johnson's withdrawal from the presidency.

Bush and Conant never had much occasion to reveal their racial politics during their years of public prominence. But Conant's biographer James Hershberg recounts how Conant refused to protest the racial exclusiveness of one of his clubs in the 1960s. Bush, who in his retirement became somewhat more cranky and uninhibited than he had been during his prime, was far more expressive. He divested himself of a racist tirade to a journalist in the 1960s, denouncing the *Brown v. Board of Education* decision.[9] The Court made what he considered to be a "colossal mistake," and he went on to argue that blacks wanted (and were entitled to) legal equality but did not want, and could never attain, social or intellectual equality with whites. "Many are children in their attitudes," he said of black Americans, "and I believe that burning intellectual ambition is comparatively rare among them, although there are notable exceptions." When friends wrote to chide him and offer him the time-honored excuse of being "misquoted," Bush refused to recant. "Certainly there is nothing in the article that I did not say," he wrote one correspondent, "although I said a great deal more than that."[10]

The ideological founders of America's corporate-military-industrial-academic order had nothing of use to say about the political crises in the 1960s. Their entire ideology had been designed to avoid and evade precisely such problems. American life could not be contained within either the soothing boundaries of utopian corporate science, the rational, cooperative visions of the wartime leaders, or the fearful, Spartan sterility of the scientific militarists. All left their lasting mark for better and for worse on institutional, political, and cultural life. But the fundamental American problems of race and inequality and disparities in wealth and power would remain into the 1960s and beyond. So too would important aspects of corporate science and scientific militarism. The notion of achieving technological solutions to military and political problems, for example, was evident throughout the American misadventure in Vietnam and continued into the 1980s and 1990s with American fascination with "smart bombs" in the Gulf War and with Edward Teller's notions of a "missile shield" against nuclear rockets and other weapons of mass destruction. The technical unfeasibility of such a scheme has proven no barrier to its funding and support by both Republican and Democratic politicians.

Americans of the 1970s, 1980s, and 1990s were far more cynical and mistrusting toward elites than were Americans of preceding generations, yet many in the 1980s would call for a "Manhattan Project" approach to the AIDS crisis, certain that the sheer collaborative power of elites, operating with massive amounts of money, could solve an intractable medical problem like the HIV virus. Old patterns of thought die hard, especially when they become entrenched in powerful institutional and political structures. At the end of the twentieth century, Americans were still struggling over the problems of how to use scientific expertise, how to manage or control scientific and technological change, and, perhaps most important, they were still uncertain over what they wanted to do with their prosperous, innovative technological culture. Not even scientific experts could provide the answers.

## NOTES

## Introduction

1. The seminal works in these areas include (on professionalization) Layton, *Revolt of the Engineers*; Kevles, *Physicists*; Noble, *America by Design*; Reingold, ed., *Sciences in the American Context*; McMahon, *Making of a Profession*; Reich, *Making of American Industrial Research*; Wise, *Willis R. Whitney*; Servos, *Physical Chemistry*; Israel, *From Machine Shop to Industrial Laboratory*. Apart from Kevles, all of these authors deal with the pre-1941 period, and most are concerned with delineating the different factions within the professional groups they study rather than examining how those factions participated in larger political and cultural conflicts.

On science and the military during and after World War II, see Sherry, *Preparing for the Next War* and *Rise of American Air Power*; Sherwin, *World Destroyed*; Herken's three books *Winning Weapon, Counsels of War*, and *Cardinal Choices*; Graebner, ed., *National Security*; Bundy, *Danger and Survival*; Evangelista, *Innovation and the Arms Race*; Hogan, *Cross of Iron*. Schweber, "Mutual Embrace of Science and the Military," seeks to connect the professional experiences of the 1930s with the organizational style of the postwar labs. But Schweber errs, I think, in regarding only the academic, big-science laboratories as the models for Los Alamos, the Radiation Laboratory at MIT, and other war-related labs. There were ideological influences, as well as organizational patterns, that can be traced back not just to Berkeley but to the world of the industrial research labs. Leslie's *Cold War and American Science* sees a strong interplay between military research, academic laboratories, and the business spin-offs that resulted from weapons research (although Leslie confines his analysis to the postwar era). Kleinman's *Politics on the Endless Frontier* also sees influences from the world of industrial research on the development of postwar research institutions but provides a narrower explanation than I do of the motivations of scientists in their dealings with the state. Hart, *Forged Consensus*, along with the works of Christophe Lecuyer on MIT and Rebecca Lowen on Stanford, provide good analyses of the factors shaping institutional and political change. See Lecuyer, "MIT, Progressive Reform, and 'Industrial Service'" and "The Making of a Science-Based Technological University," and Lowen, "Transforming the University" and *Creating the Cold War University*.

The three-volume government-sponsored history of the Atomic Energy Commission is an invaluable narrative guide to the events of the 1939–61 period, even though the work bears some of the limitations of an "official" history (it did not, for example, deal with any of the recently documented human radiation experiments that occurred during the 1940s and 1950s): Hewlett and Anderson, *New World*; Hewlett and Duncan, *Atomic Shield*; Hewlett and Holl, *Atoms for Peace and War*. See also Clarfield and Wiecek, *Nuclear America*, and Balogh, *Chain Reaction*.

2. Keller, *Affairs of State*; Skowronek, *Building a New American State*.

3. See Brinkley, "Writing the History of Contemporary America."

4. There is a large, if narrowly focused, literature on the internal histories of particular labs, particular fields, and even particular devices. For a sampling, see the works of Paul Forman, David DeVorkin, and Allan Needell cited in the bibliography.

Chapter 1

1. Keller, *Affairs of State*; Rodgers, *Work Ethic*; Montgomery, *Fall of the House of Labor*; Dawley, *Struggles for Justice*.

2. Degler, *In Search of Human Nature*; Frederickson, *Black Image in the White Mind*; Kevles, *In the Name of Eugenics*.

3. Charles Rosenberg noted that nineteenth-century Americans lived in a society of "unresolved and ordinarily unstated contradictions" such as the contradiction between the nation's democratic creed of equality for all and the disturbing reality that economic and political power was being aggregated into the hands of a few. See Rosenberg, "Science and American Social Thought," 141–42.

4. I am borrowing Stephen Skowronek's term from his *Building a New American State*.

5. Bender, "The Cultures of Intellectual Life," "The City and the Professions," "Science and the Culture of American Communities," and "The Erosion of Public Culture: Cities, Discourses, and Professional Disciplines," all in Bender's *Intellect and Public Life*. Ross, *Origins of American Social Science*; Kloppenberg, *Uncertain Victory*; Haskell, ed., *Authority of Experts* and *Emergence of Professional Social Science*; Furner, *Advocacy and Objectivity*.

6. Ross, "Modernist Social Science," 171–89. Ross has also made the intriguing argument that modernism, with its "sense of perpetual transition," has diverted Americans from a historical sense of social change (Ross, "Historian's View," 45).

7. See Hofstadter, *Social Darwinism in American Thought*.

8. Keller, *Affairs of State*, viii; Dawley, *Struggles for Justice*.

9. Harris, "The Gilded Age Revisited"; Harris, "Cultural Institutions," in which he notes, "Whether or not their planners and governors realized it, such institutions expressed . . . the driving needs of an economic system that required the maintenance of unequal power distributions, passively satisfied labor forces, punctual, obedient, industrious workers and ambitious, hungry, manipulable consumers" (36). The logic of cultural institutions like the museum "was conservative." They were formed to reshape aspects of modernization in the name of "traditional values" (39). See also Harris, "Museums, Merchandising and Popular Taste."

10. See Harris, *Cultural Excursions*, 56–147. For the early history of the Museum of the Peaceful Arts, and especially for the views of the museum's benefactor about its cultural and political purpose, see the Last Will and Testament of Henry Towne (1923), New York County Surrogate's Court. The court file of the litigation concerning the estate's interest in the museum provides a useful catalog of the museum's sponsors and shows the growing presence of corporate executives on the board. See also RG 2, Box 20, Folder 196, Rockefeller Foundation Archives. On Towne's ideas on labor relations, see his *The Adjustment of Wages*; "Proposed Anti-Trust Legislation"; "For the Prevention of Railroad Strikes"; "Railroad Strikes"; and "The Neutral Line."

11. George F. Kunz to Robert H. Adams, Dec. 11, 1925, RG 2, Box 20, Folder 196, Rockefeller Foundation Archives. See Lecuyer, "MIT, Progressive Reform, and 'Industrial Service.' "

12. See memorandum of Thomas B. Appleget, Sept. 16, 1926, RG 2, Box 20, Folder 196, Rockefeller Foundation Archives; Gwynne, *Museums of the New Age*, 8–9.

13. Gwynne, *Museums of the New Age*, 11–12.

14. Reich, *Making of American Industrial Research*, 136–42; see also Galambos, "Theodore N. Vail."

15. Hughes, *American Genesis*, 151–52.

16. Danielian, *AT&T*, 102–7.

17. Layton, *Revolt of the Engineers*, 55, 57, 64; Akin, in *Technocracy and the American Dream*, xiii, 5, also noted that engineers were a profession in search of an identity in the early twentieth century. See also Haber, *Authority and Honor*; Balogh, "Reorganizing the Organizational Synthesis."

18. Layton, *Revolt of the Engineers*, 67–74. See also McMahon, *Making of a Profession*; Haber, *Authority and Honor*. Noble, in *America by Design*, argues that corporate capitalism proletarianized engineers, which is an oversimplification.

19. Reich, *Making of American Industrial Research*, 151; Jewett, "Research in Industry," 197–98. For a comparison of how other corporate scientists dealt with this problem, see Hounshell and Smith's study of Du Pont's R&D program, *Science and Corporate Strategy*, 223–49, and Wise, *Willis R. Whitney*.

20. Jewett, "John Joseph Carty."

21. There was one bright side to this threat: the mechanical repeater that the company was trying to develop for its transcontinental wire system could also be used to transmit messages through the "ether." At worst, AT&T could compete in the radio-telephone industry with the most up-to-date equipment. "A successful telephone repeater," Carty told the board, "might put us in a position of control with respect to the art of wireless telephony should it turn out to be a factor of importance" (Danielian, *AT&T*, 104–5).

22. Reich, *Making of American Industrial Research*, 160, 195.

23. Douglas, *Inventing American Broadcasting*, 240–44.

24. Ibid., 248; see *New York Times*, Oct. 17, 1915, sec. 4, p. 9.

25. *New York Times*, Oct. 17, 1915, sec. 4, p. 9.

26. Ibid.

27. Ibid.

28. For a vivid example of this mentality, see Reich's discussion of Willis Whitney, Coolidge, and Langmuir in Reich, *Making of American Industrial Research*, 69–128; see also Hounshell and Smith, *Science and Corporate Strategy*, 286–317, and Wise, *Willis R. Whitney*.

29. Marchand, *Creating the Corporate Soul*; Ohmann, *Selling Culture*.

30. Quoted in Rodgers, *Work Ethic*, 222. Crunden, in *Ministers of Reform*, 68, points out that Progressives succeeded when, like Jane Addams at Hull House, they were able to "appeal to old and new tendencies in American life" and to practice what he calls "innovative nostalgia." See also Marchand, *Creating the Corporate Soul*, which is an excellent study of how older values were invoked to (literally) sell a new ideology to the American public.

31. Kloppenberg, *Uncertain Victory*; Chandler, *Visible Hand*; Sklar, *Corporate Reconstruction*.

32. Marx, *Machine in the Garden*; Kasson, *Civilizing the Machine*; Israel, *From Machine Shop to Industrial Laboratory*.

33. Bellamy, *Looking Backward*, 78. For an examination of changing cultural attitudes about technology and engineering, see Tichi, *Shifting Gears*.

34. Bellamy, *Looking Backward*, 61.

35. Ibid., 139.

36. Nye, *Electrifying America*, 147–49, points out that 160 utopian books appeared in the year following the publication of *Looking Backward*. Bellamy, of course, was also not the only American who saw corporate capitalism as an instrument of progressive evolution. Walter Rauschenbusch saw the corporate order as an "evolutionary process by which a great cooperative system will eventually be built up" (quoted in Kloppenberg, *Uncertain Victory*, 246).

37. Carty, *Extension of the Spoken Word*; "Science and the Industries"; "Science, the Soul of Prosperity."

38. Carty, "Science and the Industries."

39. Carty, "Science and Business."

40. Carty, "Science and Progress in the Industries."

41. National Academy of Sciences, *Biographical Memoirs*, 72.

42. Quoted in Rhodes, *John J. Carty*, 58, 185. See Carty, *Psychiatry and International Relations*.

43. Leach, *Land of Desire*; Lears, *Fables of Abundance*; Ohmann, *Selling Culture*.

44. Marchand, *Creating the Corporate Soul*, 48–87, 130–63. The N. W. Ayer advertising firm handled AT&T's successful campaign, but Bruce Barton used identical themes on behalf of the General Electric and General Motors companies in the 1920s. The motives of the companies for creating this publicity were also the same: to create public goodwill and, just as important, to solidify the morale and loyalty among the company's own employees, especially the management. Management loyalty was essential not only to ensure effective operations but also to act as a political buffer between the company leaders and the lower-level employees. By creating a common ideological bond between owners and managers, the managers would be likely to support the owners against unionization efforts and other threats to the owners' control of the company.

45. Steinmetz, *America and the New Epoch*, 137–41, 156, 137, 158, 35, 32. See also Kline, *Steinmetz*.

46. Tobey, *American Ideology of National Science*, 182–83, 185–90.

47. Chandler, *Visible Hand*; Wiebe, *Search for Order* and *Self-Rule*; Zunz, *Making America Corporate*.

48. See Hollinger, *In the American Province*, 51–52, 55, and James Kloppenberg's comments on Lippmann's use of science in his *Uncertain Victory*, 318–21.

49. Lippmann, *Drift and Mastery*, 50.

50. Ibid., 43.

51. Ibid., 49.

52. Ibid., 85.

53. Ibid., 148.

54. Ibid., 151, 155, 162.

55. Croly, *Promise of American Life*, 264.

56. Ibid., 416.

57. Ibid., 207.

58. Ibid., 206.

59. Hoover, *American Individualism*, 44.

60. Gilbert, *Designing the Industrial State*, 52, 74.

61. Sklar, *Corporate Reconstruction*, 110–12.

62. Kloppenberg, *Uncertain Victory*, 9.

63. Walter Lippmann provided some of the most provocative statements of this theme in his books *Public Opinion* (1922) and *The Phantom Public* (1925). The rise of expertise as a political force is chronicled in a vast literature, among the highlights of which are Wiebe's classic *The Search for Order*, which bids to be eclipsed in significance, I think, by his superb *Self-Rule*. See also Hawley, *Great War*; Lustig, *Corporate Liberalism*; Karl, *Uneasy State*; Eakins, "Origins of Corporate Liberal Policy Research"; McCartin, *Labor's Great War*; Weinstein, *Corporate Ideal*; Gilbert, *Designing the Industrial State*; Critchlow, *Brookings Institution*; Steidle, " 'Reasonable' Reform"; Alchon, *Invisible Hand of Planning*.

64. Zunz, "Producers, Brokers," 293.

65. Jordan, *Machine Age Ideology*.

66. See works such as Nye's *Image Worlds* and *Electrifying America*, along with Zunz's

*Making America Corporate*, on this theme of corporations and the public adapting themselves to each other.

67. Nye, *Electrifying America*; Nye, *American Technological Sublime*; Cowan, *More Work for Mother*; Fischer, *America Calling*; Noble, *America by Design*, xvii. Marchand, in *Creating the Corporate Soul*, shows the success of companies like AT&T and GE in getting some members of the public not just to tolerate but actually to love the formerly "soulless" corporation. See also Tobey, *Technology as Freedom*; Platt, *Electrical City*.

68. Nye attacks the "neo-Marxian" fascination with Fordism and Taylorism which some scholars have seen as typical but was not. See Nelson, *Frederick W. Taylor*; Nye, *Consuming Power*, 133–34, along with his *Electrifying America* and *American Technological Sublime*.

69. Lears's *Fables of Abundance* is especially good on the different possible "readings" of advertising and the different intellectual and even spiritual uses of the artifacts and symbols of the new consumer culture. Yet he also keeps the reader's attention on the growing strength of a corporate and professional ethos around this culture, which served to distort or stunt its possibilities.

70. Steinmetz, *America and the New Epoch*, 189–98.

71. The quoted phrases are from Wiebe's *Self-Rule*. Disturbing examples of this process can be seen in Jones, *Bad Blood*, and in the exploitative radiation experiments conducted by the AEC in the Cold War era. See Kevles, *In the Name of Eugenics*; Degler, *In Search of Human Nature*; Tucker, *Science and Politics of Racial Research*. Many of the scientific administrators I examine in the subsequent chapters were either explicitly racist or were all too willing to accommodate themselves and their institutional actions to prevailing racist sentiments.

72. For example, Carl Degler points out, in *In Search of Human Nature*, that belief in American "exceptionalism" helped drive out racism, at least from the public sphere. Brian Balogh's work is instructive here. Looking at American history through the lens of the "organizational synthesis," one sees a society that is paradoxically less democratic and yet more porous, more susceptible to public manipulation. See Balogh, "Reorganizing the Organizational Synthesis."

73. Quoted in Alchon, *Invisible Hand of Planning*, 47; Bender, *Intellect and Public Life*, 132.

74. Gruber, *Mars and Minerva*; Kennedy, *Over Here*; Cohen, *Making a New Deal*.

75. Quoted in Cochran and Miller, *Age of Enterprise*, 337. Louis Galambos noted in *Public Image of Big Business*, 221, that by 1929, the middle class "had come to accept the giant corporation as a permanent feature of their society. For the most part they had not learned to love big business but they decided they could live with it."

76. Marchand, *Creating the Corporate Soul*. On the continuing tension between workers and employers in the science-based industries, see Schatz, *Electrical Worker*.


## Chapter 2

1. Jewett, "Research in Industry," 196.

2. Ibid., 197, 199–200. See also Jewett's 1932 manuscript, "Philosophy and Practical Application," 35–36.

3. Reich, *Making of American Industrial Research*, 205–17.

4. See Wise, "New Role for Professional Scientists," 410; Reich, "Irving Langmuir," 200; Conant, *My Several Lives*, 25.

5. Whitney was probably speaking about W. D. Coolidge.

6. Hounshell and Smith (*Science and Corporate Strategy*, 299) ignore this passage when they describe Conant as exemplifying the typical academician's snobbery toward corporate science. Conant may have expressed reservations about his Du Pont consultancy, which he held for only a brief time, but that does not mean he objected to or resented the work of corporate science per se. Conant resigned the consultancy because he was elected scientific director of the Rockefeller Institute for Medical Research. See E. K. Bolton to Conant, Apr. 26, 1930, Incoming Correspondence, Box 6, Folder 1930, Conant Papers. Conant's correspondence during his years as a working scientist in the 1920s and early 1930s provides several examples of his preference for academic over industrial research. He believed first-rate students should pursue the former, while the mediocre should settle for the latter (see Conant to F. D. Holmes, Nov. 17, 1926, Outgoing Correspondence, Box 2, Folder 1926, ibid.). But these prejudices are not the whole story. His correspondence reveals his heavy involvement with industrial research both on mundane levels—purchasing supplies for experiments—and on substantive matters such as assisting in job searches for corporate research directors (see Conant to Donald K. David, Oct. 4, 1927, Folder 1927, ibid.). In a December 22, 1926, letter from Conant to Frank C. Whitmore, Conant makes clear that he objected to having organic chemists too closely associated with the "technicists," not because he felt industrial research was contemptible but because it would not help organic chemists enhance their status relative to that of the "aristocracy of physical chemistry." Conant did not, however, object to the idea of organic chemists pressing the "biochemical implications" of their work and its effect on medicine—and pharmacology. The relationship between organic chemistry and medicine was, he believed, "the most profitable alliance for the future of organic chemistry as a science in this country" (see Box 2, Folder 1926, ibid.).

7. Servos, *Physical Chemistry*, 207.

8. See Hounshell and Smith, *Science and Corporate Strategy*, 314–16.

9. Servos notes there were "few real examples" of the scientist who utterly disdained industrial research (*Physical Chemistry*, 109). Other physical chemists who became prominent directors of industrial research included John Johnston and R. E. Zimmerman of US Steel, Guy Buchanan of American Cyanimid, and Roger Williams, the vice-president for research at Du Pont. Servos notes that "for Noyes, Lewis, and their associates, the future of physical chemistry lay not in insular independence but in intimate interdependence" (ibid., 153–54).

10. Ibid., 219, 238, 233.

11. Ibid., 250.

12. Jewett, "Research in Industry," 198, 200.

13. Jewett, "Philosophy and Practical Application," 21–22, 24.

14. Jewett, "Dinner Address," Bell System Educational Conference (1924), quoted in Noble, *America by Design*, 178.

15. See LaFollette, *Making Science Our Own*, esp. chaps. 3–10.

16. Jewett, "Scientist and Engineer as Citizen," 4.

17. Ibid., 5–6.

18. Ibid., 6.

19. AT&T, *The World behind Your Telephone* (1936), New York Public Library. Perhaps one testimonial to the effectiveness of this rhetoric was the high regard in which Howard Scott of the technocracy movement held AT&T; he considered it the model of what an industrial corporation should be in a technocratic society; see Akin, *Technocracy and the American Dream*, 86. See also Galambos, "Theodore N. Vail"; Fischer, *America Calling*; Marchand, *Creating the Corporate Soul*. Marchand gave a full-length analysis of the

corporate rhetoric I quote here. He found some differences between companies in employing this trope: some, like GE and GM, would use specific scientists in their ads and public relations material if those scientists were colorful or popular figures like Steinmetz or Charles Kettering. But generally Marchand's work substantiates the point I am making: corporate advertising and publicity were used to make the monopolistic company seem a natural entity that smoothly coexisted with and even embodied American's fundamental cultural and political values. Corporations even went so far as to ape the populist imagery employed by *Life* magazine and by the leading documentary realist photographic artists of the decade. Marchand makes a particularly interesting point when he notes that the tones of all these ads became increasingly abstract and idealized over the course of the 1910s, 1920s, and 1930s. See Marchand, *Creating the Corporate Soul*, chaps. 5–6.

20. Page, *Bell Telephone System*, 166–67.

21. Ibid., 97. Note the fundamental difference between this corporatist conception of Progressive interdependence and the Deweyan sense of interdependence in which democracy was the method that would knit individual and collective goals. The propagandists of corporate science had made Progressive pragmatism their own—and fundamentally altered its meaning. Corporate production had replaced public intelligence as the driving force of social change.

22. Ibid.

23. See Nye, *Image Worlds*; Marchand and Smith, "Corporate Science on Display," 148–82. General Motors's Alfred Sloan and Charles Kettering created a "Parade of Progress" during the 1930s, a traveling caravan pageant of technical achievement, to counteract what Sloan described as "the thinking of men who believe that progress has ceased, that we must live by dividing up available jobs and accept a lower standard of living." See Marchand and Smith, "Corporate Science on Display," 153. GE created its "House of Magic" radio shows to popularize its labs and their products; it also used the ad slogan "progress is our most important product" (ibid., 160–64, 171). See also Marchand, *Creating the Corporate Soul*; Jordan, " 'Society Improved the Way You Can Improve a Dynamo,' " 69.

24. Kettering and Whitney quotes from Ross, ed., *Profitable Practice*, 57, 256. Whitney's contribution to the Ross volume was coauthored by L. A. Hawkins.

25. Memorandum by John Mills, "Advertising Bell Labs" (July 1943), Location 720502, AT&T Archives, Warren, N.J.; Willis Whitney at GE was appalled that his firm put more money into advertising its products than into the scientific research needed to create those products; see Reich, *Making of American Industrial Research*, 94.

26. Wise, *Willis R. Whitney*, 295.

27. Jewett, "Thirty-Five Years of Applied Science," 1–2.

28. Ibid., 3–4. He made similar points in his chapter "Philosophy and Practical Application," in Ross, ed., *Profitable Practice*, when he wrote (13–14) that research would not bring "unalloyed blessings" and would require "suitable social controls" as long "as man continues to be a greedy, and frequently a dishonestly greedy, creature."

29. Jewett, "Thirty-Five Years of Applied Science," 4.

30. Ibid., 14–15.

31. Rydell, " 'Fan Dance of Science,' " 527.

32. Ibid., 529–31.

33. Ibid., 535–41.

34. Jewett to F. B. Pratt, July 2, 1941, RG 2, Box 20, Folder 196, Rockefeller Foundation Archives. In 1942, time ran out on the museum. The trustees exercised the provisions in the Towne will which called for the dispersal of the $15 million bequest to the Museum of

Natural History and to the Metropolitan Museum of Art if the science museum failed to get another benefactor. The money was disbursed; the museum struggled on for a few more years and then closed its doors in 1949.

35. Jewett, "Social Effects of Modern Science"; "Place of the Science Museum," 2.

36. Jewett, "Place of the Science Museum," 5.

37. Jewett, "Application of Science to Industry," 223; "Place of the Science Museum," 2.

38. Jewett, "Science Increases Employment."

39. Dupree, *Science and the Federal Government*, 289–381.

40. Reingold, *Science American Style*, 228. See also Cochrane, *National Academy of Sciences*.

41. See Hale's "Report of the Committee on Additional Funds for Research," Nov. 5, 1925, 4, 7, Organization/Committee on Additional Funds for Research, National Academy of Sciences Archives.

42. Kevles, *Physicists*, 186–88. Only AT&T, US Steel, the National Electric Light Association, and the American Iron and Steel Institute provided corporate contributions.

43. Kargon and Hodes, "Karl Compton, Isaiah Bowman," 310; Cochrane, *National Academy of Sciences*.

44. See Kevles, *Physicists*, 252–58.

45. Dupree, *Science and the Federal Government*, 353–57; see also Kuznick, *Beyond the Laboratory*, 34–35.

46. Kargon and Hodes, "Karl Compton, Isaiah Bowman," 316.

47. Box 159, Merriam Papers.

48. Owens, "MIT and the Federal 'Angel,'" 202.

49. Quoted in Sherry, *Preparing for the Next War*, 124.

50. Kargon and Hodes, "Karl Compton, Isaiah Bowman," 312–13.

51. Compton, "Put Science to Work: A National Program," Box 159, Merriam Papers. See also *Technology Trends and National Policy, Including the Social Implications of New Inventions: Report of the Subcommittee on Technology to the National Resources Committee* (Washington, D.C.: Government Printing Office, 1937).

52. Kevles, *Physicists*, 255–58.

53. Quoted ibid., 262. See Servos, "Industrial Relations of Science"; Owens, "MIT and the Federal 'Angel.'"

54. Dupree, *Science and the Federal Government*, 348–49; *Science* 84 (Oct. 30, 1936): 393–94.

55. *Science* 84 (Oct. 30, 1936): 393–94.

56. Quoted in Kuznick, *Beyond the Laboratory*, 53–54.

57. See "Transcript of a Discussion at the Meeting of the Science Advisory Board, September 20 and 21, 1934," Box 159, Merriam Papers.

58. Ibid.

59. Ibid.

60. See Bush's "Statement for the Science Advisory Board Relative to the Proposed New Industries Committee" and his "Tentative Outline for Discussion of Proposed Scope of Committee Study," both in Minutes of Nov. 12, 1934, SAB Meeting, Box 159, Merriam Papers.

61. Kargon and Hodes, "Karl Compton, Isaiah Bowman," 317.

62. Servos, "Industrial Relations of Science"; Owens, "MIT and the Federal 'Angel.'"

63. Jewett convinced Compton to accept the presidency of MIT by arguing the need to move away from vocational training; see Geiger, *To Advance Knowledge*, 181. See also Servos, "Industrial Relations of Science," and Owens, "MIT and the Federal 'Angel.'" On MIT's history, see Lecuyer, "MIT, Progressive Reform, and 'Industrial Service,'" 279–300, and his "Making of a Science-Based Technological University." Lecuyer sees the

professional and institutional aspirations of people like Jewett, Bush, Karl Compton, and James Killian as distinct factors in explaining the historical changes at MIT. These leaders and the institute's faculty collaborated with but were not totally controlled by either corporate or military patrons. See also Carlson, "Academic Entrepreneurship," 536–37.

64. Owens, "MIT and the Federal 'Angel,' " 189–92.

65. Ibid., 200–211.

66. See Kuznick, *Beyond the Laboratory*, 94–95 and chap. 8.

67. Bush, *Pieces of the Action*; Zachary, *Endless Frontier*.

68. Bush, *Pieces of the Action*, 253.

69. Owens, "MIT and the Federal 'Angel,' " 193–94; Bush, *Pieces of the Action*, 255.

70. Jackson sought to justify the company's rates by grossly overvaluating the firm's assets and by dishonestly inflating the amount of equipment. In one egregious example, he included sections of pipe as items in the company's inventory of electrical poles. Cooke made a minor crusade of exposing the fraudulent work of Jackson and other pro-utility engineer-consultants. See also Layton, *Revolt of the Engineers*, 88, 162–63; McMahon, *Making of a Profession*.

71. Quoted in Layton, *Revolt of the Engineers*, 7, 242.

72. For an interesting analysis of Bush's place in the formation of New England's high-tech industry, see Warner, *Province of Reason*, 181–208. Some of Warner's conclusions about Bush's political views are, however, off the mark. He sees only a harsh, anti-democratic elitism in Bush's thought but misses the significant contradictions.

73. See E&IR: Reorganization of the Division, Proposed (1937), National Academy of Sciences Archives.

74. Bush to R. G. Harrison, Feb. 23, 1938, ibid.

75. Bush, "Office Memorandum No. 470—(Barrows)—2/1/38," ibid.

76. Bush, *Pieces of the Action*, 274.

77. Ibid., 274. See also Bush's account of some scrapes he had with the patrician Henry Stimson—a man Bush respected far more than Cameron Forbes, however (ibid., 283–85). See also a revealing comment Bush made in his retirement to James Conant, explaining why he was reluctant to write a memoir: "Very early in life, perhaps because I was a commoner among Brahmins, I became leary [*sic*] of chaps that went around tooting personal horns, and I could never bring myself to write memoirs or autobiography for that reason" (Bush to Conant, July 13, 1967, Box 18, Bush Collection, MIT Archives).

78. Bush, *Pieces of the Action*, 274.

79. Ibid., 275–76.

80. Quoted in Layton, *Revolt of the Engineers*, 191–92, 190. Hoover, *American Individualism*, 20–23, 8–9. See also Hawley, "Herbert Hoover and American Corporatism"; Karl, "Presidential Planning and Social Science Research"; Wilson, *Herbert Hoover*; Zeiger, "Herbert Hoover and the Wage-Earner."

81. Bush, "The Qualities of a Profession," Box 174, Folder: Speeches, 1937, 1939, Bush Papers, LC.

82. Ibid., 3.

83. Ibid., 5. See Bush, "Opportunity for the Professions," *Technology Review*, July 1939.

84. Hershberg, *James B. Conant*, 85–86.

85. Ibid., 89.

86. Ibid., 119, 123, 126.

87. Ibid., 39.

88. Conant, *My Several Lives*, 43–45, 51; Hershberg, *James B. Conant*, 39.

89. Hershberg, *James B. Conant*, 45–48.

90. Conant, *My Several Lives*, 49.

91. Ibid., 51–52; Hershberg, *James B. Conant*, 56.

92. Heilbron and Seidel, *Lawrence and His Laboratory*, 254.

93. Kevles, *Physicists*, 284–86.

94. Ibid., 272; Davis, *Lawrence and Oppenheimer*, 68.

95. Davis, *Lawrence and Oppenheimer*, 76–77.

96. Heilbron and Seidel, *Lawrence and His Laboratory*, 219.

97. Ibid., 239; Tuve quoted in Angelo Baracca, " 'Big Science' vs. 'Little Science,' " 382. Tuve ultimately left physics. It had, he said, "changed from a sport into a business" (ibid., 387).

98. Heilbron and Seidel, *Lawrence and His Laboratory*, 239–42, 44.

99. Jewett and King, "Engineering Progress and the Social Order," 368. Jewett was quoting Henry Adams.

100. Ibid., 368, 371.

101. Ibid., 370. See Tobey, *American Ideology of National Science*.

102. Pursell, "Science Agencies in World War II," 360.

103. Quoted in Sherry, *Preparing for the Next War*, 126.

## Chapter 3

1. Crunden, *Ministers of Reform*; Dupree, *Science and the Federal Government*; Gruber, *Mars and Minerva*, 108; Skowronek, *Building a New American State*; Karsten, "Armed Progressives," 222; Kennedy, *Over Here*.

2. Karsten, "Armed Progressives," 227. Business reformers were also conspicuously present in the preparedness movement. Howard Coffin, an engineer and vice-president of the Hudson Motor Car Company, was a key organizer for preparedness and chaired a committee of the Naval Consulting Board. John Carty recruited AT&T executive Walter Gifford to work with Coffin's organization. Hollis Godfrey, president of Drexel Institute in 1913 and a Taylorite, was also a leader in the preparedness movement. The term "business reformers" is taken from Wiebe's *Businessmen and Reform*. See Cuff, "Cooperative Impulse and War," 234–39.

3. Bush, *Modern Arms and Free Men*, 19; see also Bush, *Pieces of the Action*, 53.

4. Bush, *Pieces of the Action*, 58; Compton, *Atomic Quest*, 29–30.

5. Bush to Jewett, May 2, 1940, Box 55, Folder 1375, Bush Papers, LC.

6. Bush, *Pieces of the Action*, 40.

7. Quoted in Rhodes, *Making of the Atomic Bomb*, 362; see also Compton, *Atomic Quest*, 47.

8. Conant, *My Several Lives*, 278.

9. K. T. Compton to Bush, Mar. 17, 1941, Bush-Conant File, National Archives.

10. Bush to Jewett, June 7, 1941, ibid.

11. Ibid.

12. Smyth, *Atomic Energy*, 51. The late Stanley Goldberg offered a fascinating and subtle argument about Bush's actions during the spring of 1941 that differs from my own interpretation ("Inventing a Climate of Opinion"). Goldberg believed that Bush became convinced of the feasibility of the fission weapon sometime in the spring of 1941, shortly after receiving the March 17, 1941, letter from Karl Compton. He further argued that Bush then used the three NAS committees to push aside opposition to the program, especially the opposition of Conant and Jewett, and thereby create the appearance if not the reality of a consensus. While I agree with Goldberg that Bush was strongly motivated

by fear of a German bomb project and that he was instrumental in getting the U.S. project going, I disagree with his interpretation of Bush's motives and tactics. My objections are both evidentiary and methodological. Bush's correspondence with Jewett throughout the course of the three NAS panels reveals his uncertainty and frustration about the idea of a fission weapon. Goldberg, if I am reading him correctly, is arguing that such correspondence is not what Bush really meant, that he was placating Jewett and Conant. He wants us to draw opposite meanings from the documentary evidence, based on an inference that Bush supported the fission project but dared not push it too hard. There does not seem to be enough evidence to justify such a departure from the plain meaning of Bush's letters. I think Bush was unsure throughout the spring of 1941 for a more documentable reason: he wanted to preserve his credibility as a scientific administrator and by extension the credibility of scientists in government, and he wanted to protect his country against the threat of a German bomb. These two goals were at times contradictory, and they explain, I believe, why Bush was moving so carefully to create a professional consensus in the spring and summer of 1941. I believe he wanted a consensus that reflected his conception of how scientists should work with the state. At some point, of course, he did become an unequivocal supporter of the fission idea, but it is hard to know exactly when that happened. I believe it was the Maud draft and final reports that did the trick.

13. Quoted in Rhodes, *Making of the Atomic Bomb*, 366.

14. Jewett to R. A. Millikan, May 26, 1941, Bush-Conant File, National Archives; Millikan to Jewett, May 31, 1941, ibid.

15. Rhodes, *Making of the Atomic Bomb*, 366.

16. Compton, *Atomic Quest*, 48; Bush to Jewett, July 9, 1941, Bush-Conant File, National Archives.

17. See Bundy, *Danger and Survival*, 44–45; see also Bush's progress report to Roosevelt, dated March 9, 1942, and Roosevelt's handwritten response on March 11, 1942: "The whole thing should be *pushed* not only in regard to development, but also with due regard to time. This is very much of the essence" (emphasis in original). Both are in Bush-Conant File, National Archives.

18. Bush to Jewett, Nov. 11, 1941, Bush-Conant File, National Archives.

19. Ibid.

20. Jewett to Bush, Nov. 3, 1941, ibid.

21. Bush to Jewett, Nov. 4, 1941, ibid.

22. Bush, *Pieces of the Action*, 60.

23. Bush to Conant, Conant to Bush, undated handwritten notes, Bush-Conant File, National Archives.

24. Bush to Frank Aydelotte, Dec. 15, 1941; Harold Urey to Bush, Jan. 12, 1942, ibid.

25. K. T. Compton, *Scientists Face the World of 1942*, 23–24.

26. "The Situation on New Weapons," memo from Bush to Harvey Bundy, n.d., but with a penciled notation of Jan. 26, 1942, Box 17, Folder 389, Bush Papers, LC.

27. Bush, *Pieces of the Action*, 89, 91; see also Kevles, *Physicists*, 312–15. Furer quoted in Sherry, *Preparing for the Next War*, 135.

28. Compton, *Atomic Quest*, 114, 108–9; see also Price, "Roots of Dissent."

29. Hewlett and Anderson, *New World*, 107–10.

30. Conant to A. H. Compton, Dec. 1, 1942, Bush-Conant File, National Archives.

31. A. H. Compton to Conant, Nov. 23, 1942, ibid.

32. See Compton, *Atomic Quest*, 132–36.

33. A. H. Compton to Conant, Nov. 23, 1942, Bush-Conant File, National Archives.

34. Ibid.

35. Ibid.

36. Hewlett and Anderson, *New World*, 110–13; Compton, *Atomic Quest*, 144.

37. Hounshell, "Du Pont and the Management of Large Scale Research and Development," 248–49.

38. Hewlett and Anderson, *New World*, 120.

39. See memorandum, Feb. 26, 1942, from Thayer to Bush, Bush-Conant File, National Archives.

40. Bush to Thayer, Feb. 28, 1942, ibid.

41. Bush to Conant, Feb. 24, 1942, ibid.

42. A. H. Compton to Conant, Sept. 18, 1942, ibid.

43. Conant to A. H. Compton, Nov. 4, 1942, and A. H. Compton to Conant, Nov. 10, 1942, ibid.

44. The Urey incident is discussed in Conant's July 31, 1943, memorandum to Bush, ibid.; see also Lowen to FDR, Oct. 29, 1943, ibid. See Szilard to Bush, May 26, 1942, Jan. 14, 1944, Bush to Szilard, Jan. 18, 1944, ibid.

45. See A. H. Compton's July 23, 1943, memorandum to Wigner, ibid.

46. Ibid.

47. A. H. Compton to Wigner, July 23, 1943, Bush-Conant File, National Archives.

48. The report summary is reprinted as an appendix to Smith's *A Peril and a Hope*, 561. See also Price, "Roots of Dissent."

49. Smith, *A Peril and a Hope*, 561–63.

50. Ibid., 563.

51. Ibid., 566–67.

52. Ibid., 568, 566.

53. Ibid., 566, 568, 570–71.

54. Bundy, *Danger and Survival*, 70–71.

55. Sherwin, *A World Destroyed*, 212–13, points out that Conant also endorsed the idea that the bomb "must be used" so as to deter future wars.

56. Bundy, *Danger and Survival*, 68–70.

57. Quoted in Herken, *Cardinal Choices*, 25–29.

58. Ibid., 26. A. K. Smith argues that Oppenheimer probably drafted his report without the Franck Report before him and may not even have read it yet. But A. H. Compton noted that he made the ideas of the report known to the other scientific advisory panelists. See Smith, *A Peril and a Hope*, 50–51.

59. See Sherry, *Preparing for the Next War*, chap. 5. Brian Balogh also addresses the desire of scientists to achieve greater cultural status in his "Reorganizing the Organizational Synthesis" and *Chain Reaction*.

60. See Jewett's testimony on January 29, 1945, before the House Select Committee on Post-War Military Policy, p. 9, and before the House Committee on Military Affairs, Research and Development, May 23, 1945, p. 51; copies are in File 50.82, Jewett Papers, NAS Archives. On the differences (and the similarities) between Bush and Jewett, see also Reingold, "Vannevar Bush's New Deal for Research"; Kevles, "National Science Foundation"; and Kevles, "Scientists, the Military, and the Control of Postwar Defense Research."

61. Jewett, May 23, 1945, testimony, 58, 45.

62. Jewett, Jan. 29, 1945, testimony, 11.

63. Jewett, statement before Senate Committee on Military Affairs, Subcommittee on Technological Mobilization, Nov. 20, 1942, 196, copy in File 50.271, Jewett Papers, NAS Archives. Jewett's private correspondence about the Kilgore bill reveals the depth of his hatred of the New Deal—of which he saw Kilgore as a prime example. The senator's bill, he told Oliver Buckley in a January 5, 1943, letter, would create a "veritable American Gestapo." In a February 23, 1943, letter to Walter Gifford he also deplored the "Gestapo

philosophy" which the bill supposedly represented. Both in File 50.271, Jewett Papers, NAS Archives.

64. Bush, "Statement of Dr. Vannevar Bush, Director of Office of Scientific Research and Development, before the Select Committee on Post-War Military Policy on January 26, 1945," 2, in File 50.82, Jewett Papers, NAS Archives.

65. Ibid., 14–15, 19.

66. Jewett to Wilson, Sept. 12, 1944, Box 56, Folder 1377, Bush Papers, LC.

67. Bush, Jan. 26, 1945, statement, 8–12.

68. Bush, "Statement of Dr. Vannevar Bush," *Science* 98 (Dec. 31, 1943): 571–77, 572.

69. Jewett testimony, Jan. 29, 1945, 10. On the political and professional battles of postwar science, see Kleinman, *Politics on the Endless Frontier*; Reingold, "Choosing the Future"; Wang, "Liberals, the Progressive Left"; and Kevles's two articles "National Science Foundation" and "Scientists, the Military, and the Control of Postwar Defense Research."

70. Hewlett and Anderson, *New World*, 412, 409.

71. Smith, *A Peril and a Hope*, 251–52.

72. Bush to F. Alexander Magoun, Dec. 15, 1945, Box 68, Folder 1682, Bush Papers, LC.

73. Bush to Price, Dec. 30, 1953, Box 94, Folder 2147, ibid. See also Bush, *Pieces of the Action*, 130–31.

74. Bush, *Pieces of the Action*, 304.

75. Hewlett and Anderson, *New World*, 461–69.

76. Hershberg, *James B. Conant*, 307.

77. Hewlett and Duncan, *Atomic Shield*, 17–18; Herken, *Cardinal Choices*, 31.

78. Berkner to Chair of House Military Operations Sub-Committee, June 29, 1954, Box 11, Folder 247, Bush Papers, LC.

79. Bush to Jewett, May 7, 1946, Box 56, Folder 1377, ibid.

80. Sherry, *Preparing for the Next War*, 134.

81. Bush to Price, Dec. 30, 1953, Box 94, Folder 2147, Bush Papers, LC.

82. Conant to Bush, Jan. 3, 1949, Box 27, Folder 614, ibid.; Kevles, *Physicists*, 363.


# Chapter 4

1. Bush, *Endless Horizons*, 148, 41.

2. Bush, *Science—The Endless Frontier*, 5, 18, 7–8.

3. Bush, *Modern Arms and Free Men*, 90.

4. Ibid., 6–7.

5. Ibid., 3–4. This language strongly echoes the words of Bush's mentor, Dugald Jackson, who argued that engineers must be trained so that they "know men and the affairs of men—which is sociology; and they must be acquainted with the business methods and the affairs of the business world" (quoted in Lecuyer, "MIT, Progressive Reform, and 'Industrial Service,' " 62).

6. See below for discussion of Lilienthal's work.

7. See Alchon, *Invisible Hand of Planning*, and Wiebe, on this concept of a "deal" between elites and the public, in *Self-Rule*. Compare these works with Balogh's conception of the relation between experts and the public in *Chain Reaction*.

8. Zunz, *Making America Corporate*; Marchand, *Creating the Corporate Soul*; Geiger, *To Advance Knowledge* and *Research and Relevant Knowledge*; Levine, *American College and the Culture of Aspiration*; Livingston, *Pragmatism and the Political Economy of Cultural Revolution*.

9. See Hart, *Forged Consensus*. Hart sees science policy being fought out in the political and legislative arena and does not address, directly, the relationship between experts and the public. Bush's battles with Kilgore have been documented in Kevles, *Physicists*, and in Kleinman, *Politics on the Endless Frontier*. For Bush's impact on the transformation of higher education, see especially Geiger, "Science, Universities, and National Defense," and Lowen, *Creating the Cold War University*.

10. Bush, *Modern Arms and Free Men*, 186.

11. Ibid., 217–18.

12. Ibid., 260–61. Bush is here adopting the argument of Walter Lippmann in *The Phantom Public*: there is a small group of key individuals who, by dint of their expertise and institutional position, are making the key decisions and taking the key actions in American society. But whereas Lippmann saw the public's role as limited to that of exercising a negative check on elites by expressing electoral disapproval when things went awry and selecting different elites to fix those problems, Bush sees the elites as "distilling" the public's will.

13. Ibid., 263.

14. For Bush's and Conant's roles in the Committee on the Present Danger, see Sanders, *Peddlers of Crisis*, esp. chaps. 1–3. Sanders seems more certain of Bush's political perfidy than he does of his professional identity: he refers to him as "an atomic scientist" (54) and even as a "renowned nuclear physicist" (93).

15. Bush to Conant, Oct. 20, 1945, Box 27, Folder 614, Bush Papers, LC.

16. Address of Vannevar Bush at Washington University, Feb. 22, 1946, Box 127, Folder 3116, ibid.

17. Compton, *Atomic Quest*, 280–81; Boyer, *By the Bomb's Early Light*, 140.

18. Compton, *Atomic Quest*, 311.

19. Ibid., 334–35, 348.

20. James Conant, "Mobilizing American Youth," *Atlantic* 170 (July 1942): 50.

21. Ibid., 51–53.

22. James Conant, "Wanted American Radicals," *Atlantic* 171 (May 1943): 41–45.

23. Ibid., 43.

24. Quoted in Hershberg, *James B. Conant*, 292–94. See also Bernstein, "Seizing the Contested Terrain of Early Nuclear History."

25. Hershberg, *James B. Conant*, 294–301.

26. Ibid., 390; see chap. 20, generally.

27. Ibid., 419.

28. Ibid., 436; see also 434–38.

29. Ibid., 430–33, 460–61. See also Wang, *American Science in an Age of Anxiety*.

30. Conant, *Education in a Divided World*, 119.

31. Conant, *Science and Common Sense*, 17.

32. Conant, *Education in a Divided World*, 119–20; *Science and Common Sense*, 10.

33. Conant, *Education in a Divided World*, 121.

34. Ibid., 120–21.

35. Ibid., 33, 4, 59.

36. Lilienthal, *TVA*, 2–3.

37. Ibid., 3–4, 77–78.

38. Ibid. 106; "don't do it again," 113.

39. Ibid., 106, 48, 33.

40. Ibid., 118.

41. Ibid., 57, 74.

42. Ibid., 48, 179; Neuse, *David E. Lilienthal*; McCraw, *TVA and the Power Fight*; Talbert, *FDR's Utopian*.

43. Lilienthal, *TVA*, 184–85, 99, 97.

44. Ibid., 66, 69.

45. Lilienthal, *Journals*, 2:15, 28. Lilienthal was not the only person who saw the connection between atomic energy regulation and the "resource management" practiced by the TVA. James R. Newman and Byron S. Miller, two of the principal drafters of the AEC legislation, wrote in 1946 that they hoped not only that the AEC would carry on the New Deal legacy of the TVA but that it would lead to changes in "the structure of society itself." They also made what seems now to be an ominous point: that there was never any discussion in the Congress during the AEC debates as to the constitutional authority for this action. It was simply assumed to be another exercise of the extraordinary powers that had been exercised during wartime—and were now to be exercised in peacetime as well. See Newman and Miller, *Control of Atomic Energy*, 4, 21, 22.

46. My references are to a magazine form of the report, published by Doubleday in 1946, a copy of which is contained in Oppenheimer's Papers in the Library of Congress, Box 191.

47. Ibid., 28.

48. Ibid., 28–29.

49. Lilienthal, *Journals*, 2:30.

50. Ibid., 42. See also Holloway, *Stalin and the Bomb*.

51. Lilienthal, *Journals*, 2:39–40, 453, 462, 465, 488–89, 361, 444.

52. Lilienthal, *Change, Hope, and the Bomb*, 29; Hewlett and Duncan, *Atomic Shield*, 89.

53. Lilienthal, *Journals*, 2:233, 357, 360–61, 531. See also Wang, *American Science in an Age of Anxiety*.

54. Lilienthal, *Journals*, 2:532, 534, 543, 562.

55. Hewlett and Duncan, *Atomic Shield*, 324, 334.

56. Atomic Energy Commission, *In the Matter of J. Robert Oppenheimer*, 69.

57. Lilienthal, *Journals*, 2:229, 263; Boyer, *By the Bomb's Early Light*, 294–95.

58. See Oppenheimer to Lilienthal, June 18, 1948, Box 46, Oppenheimer Papers.

59. Ibid.

60. Lilienthal, *Journals*, 2:250, 134–35, 311.

61. Ibid., 234–35.

62. Pfau, *No Sacrifice Too Great*, 7–17.

63. On Strauss's early career as an investment banker, see ibid., 26–40.

64. Lilienthal, *Journals*, 2:234–35, 238–40.

65. Ibid., 383, 385.

66. Ibid., 383, 385, 399, 445, 469, 568.

67. Ibid., 569.

68. Strauss, *Men and Decisions*, 216–17.

69. Lilienthal, *Journals*, 2:582, 591.

70. A copy of Oppenheimer's statement, as well as the majority and minority reports, is in Cantenlon et al., eds., *American Atom*, 121–23. See also Sylves, *Nuclear Oracles*.

71. Quoted in Hershberg, " 'Over My Dead Body,' " 398.

72. Strauss, *Men and Decisions*, 219.

73. Quoted in Herken, *Counsels of War*, 56.

74. Lilienthal, *Journals*, 2:594, 601. See also Bundy, *Danger and Survival*, 209–14.

75. Bundy, *Danger and Survival*, 211–12; Hewlett and Duncan, *Atomic Shield*, 401; Lilienthal, *Journals*, 2:510.

76. Lilienthal, *Journals*, 2:501, 627.

77. Ibid., 630, 632–33.

78. Hershberg, *James B. Conant*, 481–82; Hershberg, " 'Over My Dead Body,' " 405.

79. Lilienthal, *Journals*, 2:577.

80. Hewlett and Duncan, *Atomic Shield*, 445; Lilienthal, *Journals*, 2:593, 598; Herken, *Winning Weapon*, 320.

81. Lilienthal, *Journals*, 2:67, 278–79. See also the interesting passage in an essay of Lilienthal's defending the TVA in 1946: "The perennial speculation, 'Is America moving to the Right or the Left?' has little meaning in these days." He saw political life divided instead between "the vigorous, the imaginative, the confident" and the "tired, the complacent, the fearful" ("TVA: An American Invention," *Atlantic Monthly* 177 [Jan. 1946]: 105–9).

82. The political effects of the new wartime organizations are discussed in Balogh, *Chain Reaction*; Hart, *Forged Consensus*; and Price, *Government and Science*.


## Chapter 5

1. For examples from this large literature, see the works of Needell, DeVorkin, Leslie, Seidel, and Forman cited in the bibliography.

2. Forman, "Behind Quantum Electronics," 206.

3. See Michael Hogan's *Cross of Iron* for an extended treatment of the arguments over the dangers of a "garrison state."

4. This deterministic approach has been applied to the study of prewar science as well. See Noble, *America by Design*. The term "reserve labor pool" is from Mukerji, *Fragile Power*, a book that tries, unsuccessfully I believe, to explain the political relations between all scientists and the state based on a study of oceanographers, none of whom are identified by name.

5. Paul Forman argues that "military agencies and consultation on military problems had effectively rotated the orientation of academic physics towards techniques and applications." The idea of scientists laboring under a "false consciousness" is a major theme of Forman's essays on this subject. Scientists, he writes, "pretended a fundamental character to their work that it scarcely had." They enjoyed only the "illusion of autonomy," and they were calling the tune far less than they thought ("Behind Quantum Electronics," 228–29).

6. England, *Patron for Pure Science*, 154, 142, 57.

7. Pitzer to Shenstone, Mar. 22, 1951, Reel 48, Lawrence Papers. The AEC's research programs were organized, Pitzer noted, around three goals: increasing the general fund of basic information about atomic energy, continuing the training of qualified students, and interesting the qualified scientific personnel of the country in the problems of the AEC (Pitzer to E. O. Lawrence, Sept. 27, 1949, ibid.).

8. Thomas H. Johnson to E. O. Lawrence, June 7, Sept. 9, 1946, Reel 14, ibid.

9. Address to Northwestern Institute of Technology, Sept. 8, 1955, ibid.

10. Quoted in Needell, "From Military Research to Big Science," 293.

11. Quoted in Forman, "Behind Quantum Electronics," 151, 225 n. 139.

12. Murray to Ford, Dec. 14, 1950, Reel 48, Lawrence Papers.

13. Seidel, "Home for Big Science," 164, 166. The national laboratories justified their existence, Seidel noted, "in the nature of the organization of work in the laboratories." Only these labs "combined talents of many disciplines," and only they were able to carry out the large-scale programs required to meet the government's needs. This argument was effective in sustaining the laboratories. "The Commission accepted the argument that it was not what the laboratories did but how they did it that was important." The

national labs would carry out the "multidisciplinary large scale research that was too risky for profit-makers, but not for the guardians of nuclear preparedness."

14. See Hogan, *Cross of Iron*, and Sherry, *In the Shadow of War*, on the growing assertiveness and aggressiveness of military leaders in the late 1940s. The ambitions of the military and its receptiveness to new weapons certainly played a key role in making scientific militarism possible. I am examining the cultural and intellectual basis upon which militarism was able to gain legitimacy in American political culture in a forthcoming book, *War-Torn America*.

15. Atomic Energy Commission, *In the Matter of J. Robert Oppenheimer*, 322, 325. McGeorge Bundy drew attention to this issue of a double standard by arguing that Strauss & Co. would have readily forgiven Oppenheimer's indiscretions if he had been on their side of the Super debate (see *Danger and Survival*, 315).

16. Teller, with Brown, *Legacy of Hiroshima*, 8, 10.

17. Ibid., 22–23.

18. Ibid., 42. For the very similar arguments he made eight years earlier before the Gray Board, see Atomic Energy Commission, *In the Matter of J. Robert Oppenheimer*, 716. Richard Rhodes has mounted a very compelling refutation of this argument in his history of the hydrogen bomb, *Dark Sun*. According to Rhodes, it was Teller's obsession with achieving a massive thermonuclear weapon that prevented the quicker completion of a more modest but workable Super design.

19. Teller and Brown, *Legacy of Hiroshima*, 31.

20. Ibid., 31, 45.

21. Atomic Energy Commission, *In the Matter of J. Robert Oppenheimer*, 716.

22. Ibid., 497.

23. Teller quoted in Divine, *Blowing on the Wind*, 147.

24. Gilpin, *American Scientists and Nuclear Weapons Policy*, 106.

25. Alvarez, *Alvarez*, 169, 171.

26. Atomic Energy Commission, *In the Matter of J. Robert Oppenheimer*, 776.

27. Ibid., 777–78; Alvarez, *Alvarez*, 170.

28. Atomic Energy Commission, *In the Matter of J. Robert Oppenheimer*, 805.

29. Alvarez, *Alvarez*, 170.

30. Goodchild, *J. Robert Oppenheimer*, 199. "Military establishment" was a phrase used by Oppenheimer in his 1954 testimony and quoted by Goodchild, 243.

31. Gilpin, *American Scientists and Nuclear Weapons Policy*, 110.

32. Ibid., 4.

33. Hewlett and Duncan, *Atomic Shield*, 440, 411–14, 415. Conant would come to see the campaign against Oppenheimer as indirectly targeting himself and Bush. He wrote Bush in March 1954 about "rumors that the patriotism of some or all of us involved in the [H-bomb] recommendation has been impugned" (Conant to Bush, Mar. 26, 1954, Box 4, Conant Papers. See also Hershberg, " 'Over My Dead Body' ").

34. York, *Making Weapons, Talking Peace*, 47.

35. Apr. 11, 1951, Transcript of Hearing re Production Particle Accelerators, Reel 50, Lawrence Papers. This transcript contains some unintentionally amusing exchanges between Lawrence and the committee when the latter would ask him to try to explain in laymen's terms how these accelerators work. Try as he might to keep his explanations as simple as possible, Lawrence inevitably used terms that were beyond the grasp of the members. When Lawrence spoke about neutron capture, Mr. Durham asked, "How do you hold them in storage?" Dr. Lawrence: "How is that?"

36. Teller and Brown, *Legacy of Hiroshima*, 54.

37. York, *Making Weapons, Talking Peace*, 63–65.

38. Ibid., 62–65.

39. Ibid., 66.

40. Ibid., 66–67; see also Childs, *American Genius*, 444–45.

41. Quoted in Lifton and Markusen, in their absurdly titled and just as absurdly conceived book, *The Genocidal Mentality*, 116.

42. York, *Making Weapons, Talking Peace*, 67, 70–71.

43. Ibid., 75.

44. York, Aug. 20, 1978, interview, 11, Bancroft Library, University of California, Berkeley.

45. York, *Making Weapons, Talking Peace*, 76. The lab made a commitment to fusion research early on so as to give it "some extra breadth" and to "make it more attractive in recruiting" (York, Aug. 20, 1978, interview, 1–2).

46. See Divine, *Blowing on the Wind*, 147–50; Evangelista, *Innovation and the Arms Race*; Kevles, "R&D and the Arms Race."

47. York, Aug. 20, 1978, interview, 21. He believed that Strauss's "quarrel with Oppenheimer is really an archtypical thing—it's not an accident, a single isolated event. Strauss did not tolerate people not agreeing with him" (ibid., 26).

48. Pfau, *No Sacrifice Too Great*, 93.

49. Goodchild, *J. Robert Oppenheimer*, 195.

50. Quoted in Hewlett and Duncan, *Atomic Shield*, 530–31.

51. Ibid., 536; Pfau, *No Sacrifice Too Great*, 131–32; Bernstein, "Oppenheimer Case Reconsidered," 1414.

52. Hewlett and Duncan, *Atomic Shield*, 537.

53. Herken, *Counsels of War*, 65.

54. Goodchild, *J. Robert Oppenheimer*, 213.

55. Herken, *Counsels of War*, 66.

56. Ibid., 66–67.

57. Quoted in Herken, *Cardinal Choices*, 60.

58. Atomic Energy Commission, *In the Matter of J. Robert Oppenheimer*, 562.

59. Quoted in Bundy, *Danger and Survival*, 288–89.

60. Ibid.

61. Strauss, *Men and Decisions*, 256–58.

62. Bush, "The Weapons We Need for Freedom," *Reader's Digest*, Jan. 1951, 48–51; *New York Times*, Mar. 5, 1951.

63. This evidence comes from the memory of McGeorge Bundy, who served as a secretary to the 1952 disarmament panel on which both Oppenheimer and Bush served. He reported his recollection in a letter to the editor of the *New York Times*, which was published on September 29, 1969.

64. Bush to Bradley, Apr. 13, 1950, Box 14, Folder 323, Bush Papers, LC.

65. Ibid.

66. Bush to Bradley, Feb. 23, 1951, ibid.

67. Bush to Bradley, Mar. 6, 1951, ibid.

68. Bush to Bradley, July 16, 1951, ibid. The phrase "blind spot" is in his March 6, 1951, letter to Bradley, ibid.

69. Bush wrote a memorandum that he privately circulated among Bradley and other high-ranking military officers, titled "A Few Quick," in which he called for an improved capacity to turn out small technologies rapidly. See Zachary, *Endless Frontier*, 357.

70. Bush to Bradley, Nov. 5, 1951, Box 14, Folder 323, Bush Papers, LC.

71. Bush to Bradley, Sept. 9, 1952, ibid.; see also the *New York Times*, Sept. 27, 1952, regarding the Mayo Clinic speech.

72. Vannevar Bush, "What's Wrong at the Pentagon?," *Collier's*, Dec. 28, 1952, 31–35.

73. Herken, *Cardinal Choices*, 62; Hewlett and Duncan, *Atomic Shield*, 557–58, 564, 584.

74. Millis, with Mansfield and Stein, *Arms and the State*, 379–87. The other members of the committee were Robert Lovett, Omar Bradley, Milton Eisenhower, David Sarnoff, and Arthur Flemming. The report they submitted was titled "Reorganization Plan No. 6."

75. Ibid., 382.

76. Oppenheimer, *Open Mind*, 70–71.

77. Ibid., 72.

78. Transcript of Feb. 17, 1953, Council on Foreign Relations Meeting, Box 89, Folder 2000, Oppenheimer Papers; see Strauss to Jackson, Aug. 20, 1953, Box 75, Strauss Papers.

79. Ambrose, *Eisenhower, the President*, 132, 122–23.

80. Ibid., 95.

81. Ibid., 345.

82. Pfau, *No Sacrifice Too Great*, 137–39, 145.

83. Bernstein, "Oppenheimer Case Reconsidered," 1447.


## Chapter 6

1. Atomic Energy Commission, *In the Matter of J. Robert Oppenheimer*, 385.

2. Ibid., 7, 67, 69, 18–19.

3. Ibid., 149. There was evidence in the secret file, not made available to Oppenheimer's counsel but uncovered by the AEC's official historians, which seemed to suggest that Oppenheimer was, in fact, telling the truth in 1943 to Pash and Lansdale when he said that a Soviet agent named Eltenton had asked an unnamed friend of Oppenheimer's (Chevalier) to approach three scientists on the project to see if they would be willing to pass information to the Soviets. The lies began with his "confession" to Groves that Eltenton had approached his friend Chevalier and that Chevalier had merely reported this approach to Oppenheimer, without urging Oppenheimer or anyone else to commit espionage. That was the story Oppenheimer stuck with for the rest of his life. Roger Robb never challenged the truthfulness of the confession to Groves because such a challenge threatened to drag Ernest Lawrence and Luis Alvarez into the Chevalier story. None other than George Eltenton himself, the man who approached Chevalier, told FBI agents in 1946 that Chevalier had indeed, at his request, approached three scientists: Robert Oppenheimer, Lawrence, and Alvarez. At the same time, the FBI was also interviewing Chevalier at a separate location, but agents were cross-checking the responses of the two men over the telephone. Chevalier stuck to the story of a single conversation with Oppenheimer. He did not mention three scientists. But Richard Hewlett and Jack Holl point out that there are grounds for believing the accuracy of Eltenton's story since, in the first place, it conforms with the highly detailed "cock and bull story" Oppenheimer told Captain Lansdale in 1943. In that secretly recorded interview with Lansdale, Oppenheimer stated that two of the three scientists approached by the unnamed intermediary were at Berkeley and that the Russians were especially interested in radar. Alvarez had indeed worked at the MIT radar project before coming to Los Alamos. Hewlett and Holl further speculate that the "third man" may have been Frank rather than Robert Oppenheimer. Oppenheimer's recanting of the "cock and bull story" told to Lansdale may have been an effort to protect Lawrence, Alvarez, and his brother from the attentions of the security investigators. Both Groves and Lansdale, when testifying before the Gray Board, said they thought Oppenheimer's stonewalling had been an effort to protect his brother. It was testimony which the normally tenacious Robb did not pursue. Perhaps he feared that further probing in that direction might bring out Elten-

ton's references to Lawrence and Alvarez. It may have far better suited Robb's and Strauss's purposes to nail Oppenheimer on the "lie" to which he had already admitted, and which Oppenheimer in 1954 dared not reclaim as the true story, than to prove a new lie. See Hewlett and Holl, *Atoms for Peace and War*, 97.

4. Atomic Energy Commission, *In the Matter of J. Robert Oppenheimer*, 959, 951.

5. Ibid., 1011.

6. Ibid.

7. Ibid., 1013.

8. Ibid., 1015–16.

9. Ibid., 1017, 1019.

10. Ibid., 1043, 1049, 1058.

11. Bernstein, "Oppenheimer Case Reconsidered," 1459; quoted in Hershberg, *James B. Conant*, 681. See also Hershberg's " 'Over My Dead Body.' " Hershberg argues that the Strauss-Berkeley faction also included people like Wendell Latimer, who had personal and political axes to grind against Conant as well as Oppenheimer.

12. Bush to Cooksey, May 17, 1954, Box 27, Folder 627, Bush Papers, LC.

13. Strauss to Bush, Apr. 16, 1954, Bush to Strauss, Apr. 19, 1954, Box 109, Folder 2563, ibid.

14. Bush to Strauss, Apr. 19, 1954, ibid.

15. Bush to Strauss, Apr. 28, 1954, Box 13, Strauss Papers.

16. Atomic Energy Commission, *In the Matter of J. Robert Oppenheimer*, 565, 567.

17. Vannevar Bush, "If We Alienate Our Scientists," *New York Times Magazine*, June 13, 1954, 60.

18. Ibid., 9, 62, 71.

19. Vannevar Bush, "To Make Our Security System Secure," *New York Times Magazine*, Mar. 20, 1955, 9. In October 1954, Eisenhower, when asked to respond to Bush's comments about flagging morale among scientists, said, "The scientists who have come to see me exhibit no such attitude" (England, *Patron for Pure Science*, 417).

20. See C. L. Wilson to Bush, Aug. 25, Oct. 25, Nov. 2, 22, 1954, Box 120, Folder 2918, Bush Papers, LC.

21. See Bush to Oscar Cox (memo), Apr. 16, 1951, Box 29, Folder 654, ibid.; Hewlett and Duncan, *Atomic Shield*, 466–67.

22. Lilienthal, *Journals*, 3:501.

23. Lilienthal's biographer, Stephen M. Neuse, notes that when Strauss took over the AEC, Lilienthal made a point of going to Washington to examine classified files that he thought might be used against him. When "L. gets in there," he wrote, "God knows what will happen to the files" (*David E. Lilienthal*, 234). When he tried to prepare for his 1954 testimony, he was denied access to many of the files.

24. Lilienthal, *Journals*, 3:506–7.

25. Ibid., 531.

26. Compton, *Atomic Quest*, 126. For Compton's role in the Kamen case, see Kamen, *Radiant Science, Dark Politics*, 276, 280–81.

27. Bush to Conant, Mar. 29, 1954, Box 27, Folder 614, Bush Papers, LC.

28. Bush to Price, Dec. 30, 1953, ibid.

29. Conant did meet with Eisenhower and discussed the Oppenheimer case. But the terse record of this conversation which he left in his diary suggests that, as usual, Conant was circumspect when bringing up touchy political subjects: "Saw the President for 30 min. He opened up at once on the Oppie case. Prayed it would come out O.K. but doubted it. I told him my worries about the H-bomb indecision. He hadn't heard of it!" (Conant Diary, Apr. 26, 1954, Box 10, Conant Papers). On the arms control movement and the alternative discourse of scientific politics that flourished in the 1950s, the first

stop, of course, is the *Bulletin of Atomic Scientists*. The international history of this movement is captured in Lawrence S. Wittner's two-volume work, *The Struggle against the Bomb*.

30. Daniel Kevles, "K1S2," 319–20.

31. See Rosenberg, "Origins of Overkill."

32. York, *Making Weapons, Talking Peace*, 90.

33. Quoted ibid., 90–94.

34. Neufeld, *Ballistic Missiles*, 2. After World War II work with Bush's OSRD, Gardner worked for General Tire and Rubber Company of California and then became head of an electronics firm, Hycon Manufacturing Company, in Pasadena in the early years of the military-oriented "silicon valley" industries (ibid., 96).

35. Ibid., 12. He had also argued in favor of limiting military R&D spending for fiscal year 1949 to $500 million, arguing that all basic research should be conducted by civilian agencies (ibid., 67).

36. Ibid., 149, 91.

37. Ibid., 94, 104, 133, 151. Chrysler's K. T. Keller was Truman's "missile czar," a post created after Air Force under secretary John McCone called for a Manhattan Project–style mobilization of resources behind missile development (McDougall, *Heavens and the Earth*, 105).

38. Neufeld, *Ballistic Missiles*, 147, 242–43.

39. Leslie, *Cold War and American Science*; Edwards, *Closed World*; Forman, "Behind Quantum Electronics"; Kevles, "R&D and the Arms Race"; Dockrill, *Eisenhower's New-Look National Security Policy*.

40. York, *Making Weapons, Talking Peace*, 82; Ambrose, *Eisenhower, the President*, 132.

41. Ambrose, *Eisenhower, the President*, 206, 313.

42. Ibid., 229, 184, 248, 230.

43. York, *Making Weapons, Talking Peace*, 75; Killian, *Sputnik*, 30.

44. Griffith, "Dwight D. Eisenhower and the Corporate Commonwealth," 88.

45. Quoted ibid., 91.

46. Quoted in Herken, *Cardinal Choices*, 73.

47. Ambrose, *Eisenhower, the President*, 86–87.

48. Ibid., 223–25, 396, 394.

49. Ibid., 96–97.

50. Divine, *Blowing on the Wind*, 86, 92.

51. Ibid., 148–49.

52. Quoted in York, *Making Weapons, Talking Peace*, 99.

53. Quoted in Herken, *Counsels of War*, 116.

54. Ambrose, *Eisenhower, the President*, 431–32.

55. See Rigden, *Rabi, Scientist and Citizen*, 238–39.

56. Killian, *Sputnik*, 10–11, 30, 86–87, 88.

57. Divine, *Blowing on the Wind*, 171; Killian, *Sputnik*, 110, 111, 116.

58. Killian, *Sputnik*, 112–13, xv.

59. Ibid., xvii, 6, 172.

60. Walter McDougall emphasizes that Sputnik did not lead to calls just for more scientific and technical education but for more elitism in education in general, for abandonment of the "life-adjustment" courses, and for an effort at fighting the trend toward the lowest common denominator of democratic mediocrity (*Heavens and the Earth*, 160–61). It created an ever greater impetus toward the privileging of scientific and technical knowledge in particular and expert knowledge in general. See also Clowse, *Brainpower for the Cold War*.

61. See Divine, *Blowing on the Wind*, chap. 4.

62. Ibid., 91.

63. Ibid., 42, 83, 163, 208.

64. Ibid., 170, 216.

65. Hewlett and Holl, *Atoms for Peace and War*, 400–401.

66. Ambrose, *Eisenhower, the President*, 449, 452–53, 399–400.

67. Hewlett and Holl, *Atoms for Peace and War*, 545.

68. Ibid.

69. Killian, *Sputnik*, 7, 9.

70. York, *Making Weapons, Talking Peace*, 106.

71. Ibid.; Killian, *Sputnik*, 41.

72. Quoted in Smith, *A Peril and a Hope*, 527.

73. Killian, *Sputnik*, 25, 113.

74. York, *Making Weapons, Talking Peace*, 7.

75. Ibid., 8.

76. Ibid., 25, 39.

77. Childs, *American Genius*, 267, 319–20, 267, 347.

78. Ibid., 354.

79. Ibid., 405, 286, 408–9; Heilbron and Seidel, *Lawrence and His Laboratory*, 255.

80. York, *Making Weapons, Talking Peace*, 39.

81. York, *Race to Oblivion*, 39.

82. On Lawrence's drinking, see Childs, *American Genius*, 420, 498, 446, 426.

83. Ibid., 451.

84. York, *Making Weapons, Talking Peace*, 100–101.

85. Ibid., 161–63.

86. Ibid., 96–97.

87. Ibid., 117–18.

88. Ibid., 120.

89. Ibid., 183.

90. Ibid., 184.

91. York, *Race to Oblivion*, 184–85.

92. York, *Making Weapons, Talking Peace*, 185–86; see also Herken, *Counsels of War*, 139.

93. York, *Making Weapons, Talking Peace*, 198.

94. York, *Race to Oblivion*, 21–22; see also his *Making Weapons, Talking Peace*, 199.

95. York, *Making Weapons, Talking Peace*, 114.

96. York and Wiesner, "National Security and the Nuclear Test Ban," 35.

97. York, *Making Weapons, Talking Peace*, 118–19.

98. Ambrose, *Eisenhower, the President*, 561.

99. Ibid., 514, 516.

100. See Divine, *Sputnik Challenge*, and Bundy, *Danger and Survival*.

101. Herken, *Counsels of War*, 133.

102. Ibid., 140.

## Conclusion

1. Webb, *Space Age Management*, 26, 10.

2. Ibid., 117, 15, 17.

3. Ibid., 99, 103. See Webb's Oral History Interview, Oct. 15, 1985, 20, NASM Oral History Project, available at WWW.NASM.EDU. Jerome Wiesner had disagreed with the

notion of a crash program for a moon landing. He felt research money should be used for more scientifically productive purposes. The ease with which Kennedy brushed these objections aside only proves my point about the ineffectuality of scientific critics in state science in this period. See McDougall, *Heavens and the Earth*, 378.

4. "The post-1961 NASA construction went beyond the pork barrel into the realm of social planning" (McDougall, *Heavens and the Earth*, 376).

5. Webb, Oral History Interview, Mar. 22, 29, 1985, NASM Oral History Project, at WWW.NASM.EDU.

6. See Nye, *American Technological Sublime*.

7. Iran was the principal client of Lilienthal's D&R firm, and he reimagined the Shah and his regime in terms of the benevolent state described in *TVA: Democracy on the March*. The Shah, as Lilienthal said in 1960, was "one of the most capable practical and economic leaders . . . anywhere in the world." He compared him with Lincoln and said he was "an example of a leader with a warm heart who has a concern for the people—for the poor people, for all of the people of his country." Reports that Lilienthal received in the 1970s about the murder of the regime's opponents by the secret police were peremptorily dismissed in favor of the version of reality he manufactured in his own sentimental rhetoric. See Neuse, *David E. Lilienthal*, 274–75, 308. For Johnson's fascination with this idea, see Gardner's superb *Pay Any Price*, 357.

8. Neuse, *David E. Lilienthal*, 277; Gardner, *Pay Any Price*, 357. McNamara and Dean Rusk were instructed to keep their mouths shut. Gardner's book is a model, I believe, for studying the interplay between a broad ideological context (anticommunism and containment), the dynamics of institutional momentum in the Kennedy and Johnson administrations, and the importance of particular historical actors. The complex, troubled mind of Lyndon Johnson used the ideological context and the institutional dynamics he inherited in ways that could not have been replicated by any other human being.

9. "The Supreme Court's Colossal Mistake," *Baltimore Sun*, Sept. 8, 1963. See Hershberg, *James B. Conant*, 722–27.

10. "Supreme Court's Colossal Mistake"; Bush to Charles S. Garland, Sept. 17, 1963, Box 18, Bush Collection, MIT Archives. Bush's MIT papers also contain a thirty-three-page screed against the Warren Court that was never published. See "Crisis of the Court," June 9, 1964, ibid.

# BIBLIOGRAPHY

## Primary Sources

### ARCHIVES

Vannevar Bush Collection, Institute Archives, Massachusetts Institute of Technology, Cambridge, Massachusetts.
Vannevar Bush Papers, Library of Congress, Manuscript Division, Washington, D.C.
Bush-Conant File, Office of Scientific Research and Development Collection, National Archives, Washington, D.C.
James B. Conant Papers, Nathan Pusey Library, Harvard University, Cambridge, Massachusetts.
Dwight D. Eisenhower Papers, Eisenhower Presidential Library, Abilene, Kansas.
Frank B. Jewett Collection, AT&T Archives/Records Management Services, Warren, New Jersey.
Frank B. Jewett Papers, NAS-NRC Central Policy Files, National Academy of Sciences Archives, Washington, D.C.
Ernest O. Lawrence Papers, Bancroft Library, University of California, Berkeley.
J. C. Merriam Papers, Library of Congress, Manuscript Division, Washington, D.C.
New York County Surrogate's Court, New York, New York.
J. Robert Oppenheimer Papers, Library of Congress, Manuscript Division, Washington, D.C.
Rockefeller Foundation Archives, Tarrytown, New York.
Lewis Strauss Papers, Herbert Hoover Presidential Library, West Branch, Iowa.

### CONTEMPORARY WRITINGS

Alvarez, Luis W. *Alvarez: Adventures of a Physicist*. New York: Basic Books, 1987.
Bellamy, Edward. *Looking Backward: 2000–1887*. 1988. Reprint, New York: Penguin, 1982.
Bush, Vannevar. *Endless Horizons*. Washington D.C.: Public Affairs Press, 1946.
——. *Modern Arms and Free Men: A Discussion of the Role of Science in Preserving Democracy*. New York: Simon & Schuster, 1949.
——. *Pieces of the Action*. New York: William Morrow, 1970.
——. *Science—The Endless Frontier*. 1945. Reprint, Washington, D.C.: National Science Foundation, 1990.
——. *Science Is Not Enough*. 1967. Reprint, New York: William Morrow, 1968.
Carty, John J. *The Extension of the Spoken Word*. Pamphlet. October 20, 1916. New York Public Library.
——. *Psychiatry and International Relations*. Pamphlet. January 7, 1925. New York Public Library.
——. "Science, the Soul of Prosperity." *Review of Reviews*, May 1930, 61–64.
——. "Science and Business." *Reprint and Circular Series of the National Research Council, No. 55*. Washington, D.C.: National Research Council, June 1924.
——. "Science and Progress in the Industries." *Reprint and Circular Series of the*

*National Research Council, No. 89.* Washington, D.C.: National Research Council, July 1929.

———. "Science and the Industries." *Reprint and Circular Series of the National Research Council, No. 8.* Washington, D.C.: National Research Council, February 1920.

Compton, Arthur Holly. *Atomic Quest.* New York: Oxford University Press, 1956.

Compton, Karl T. *Scientists Face the World of 1942.* New Brunswick: Rutgers University Press, 1942.

Conant, James B. *The Child, the Parent and the State.* Cambridge, Mass.: Harvard University Press, 1959.

———. *Education in a Divided World.* Cambridge, Mass.: Harvard University Press, 1948.

———. *My Several Lives: Memoirs of a Social Inventor.* New York: Harper & Row, 1970.

———. *Science and Common Sense.* 1951. Reprint, New Haven: Yale University Press, 1963.

———. *Two Modes of Thought.* New York: Trident, 1964.

Croly, Herbert. *The Promise of American Life.* 1909. Reprint, Boston: Northeastern University Press, 1989.

Gwynne, Charles T. *Museums of the New Age.* Pamphlet. 1927. New York Public Library.

Hoover, Herbert. *American Individualism.* Garden City, N.Y.: Doubleday, Page, 1922.

———. *The Challenge to Liberty.* New York: Charles Scribners & Sons, 1934.

Jewett, Frank B. "The Application of Science to Industry." *Western Society of Engineers* 38 (August 1933): 214–23.

———. "John Joseph Carty, 1861–1932." Location 720501, AT&T Archives, Warren, N.J.

———. "The Philosophy and Practical Application of Industrial Research." Manuscript. January 1932. Location 720401, AT&T Archives, Warren, N.J.

———. "The Place of the Science Museum in the Modern Industrial Community." November 1934. Record Group 2, Box 20, Folder 196, Rockefeller Foundation Archives, Tarrytown, N.Y.

———. "Research in Industry." *Scientific Monthly,* March 1939, 195–202; also in Location 720501, AT&T Archives, Warren, N.J.

———. "Science Increases Employment." *Electrical Manufacturing,* May 1934; copy in Location 550503, AT&T Archives, Warren, N.J.

———. "Scientist and Engineer as Citizen." California Institute of Technology Archives, Pasadena.

———. "The Social Effects of Modern Science." *Science,* July 8, 1932, 23–26.

———. "Thirty-Five Years of Applied Science." Location 720501, AT&T Archives, Warren, N.J.

Jewett, Frank B., with Robert W. King. "Engineering Progress and the Social Order." *Science,* October 25, 1940, 365–71.

Lilienthal, David E. *Big Business: A New Era.* New York: Harper, 1953.

———. *Change, Hope, and the Bomb.* Princeton: Princeton University Press, 1963.

———. *Journals of David Lilienthal.* Vols. 2 and 3. New York: Harper & Row, 1964, 1971.

———. *TVA: Democracy on the March.* New York: Harper & Row, 1944.

Steinmetz, Charles P. *America and the New Epoch.* New York: Harper & Brothers, 1916.

Strauss, Lewis L. *Men and Decisions.* New York: Doubleday, 1962.

Teller, Edward, with Allen Brown. *The Legacy of Hiroshima.* New York: Doubleday, 1962.

Towne, Henry R. "For Prevention of Railroad Strikes: 'Report of Committee on Public Utilities and Law,' together with 'Railroad Strikes: Their Menace and Their Lesson.' " New York: Merchants' Association of New York, 1916.

———. "The Neutral Line: A Plea for Scientific Regulation of the Tariff: Address to the National Tariff Commission at Indianapolis, February 16–18, 1909." New York: National Tariff Commission Association, 1911.

——. "Proposed Anti-Trust Legislation: Speech by Henry R. Towne before the Second Annual Meeting of the Chamber of Commerce of the United States, February 11–13, 1914." Washington, D.C.: Chamber of Commerce of the United States of America, 1914.

Towne, Henry R., Frederick A. Halsey, and Frederick W. Taylor. *The Adjustment of Wages to Efficiency.* New York: Macmillan, 1896.

York, Herbert F. *The Advisors: Oppenheimer, Teller, and the Superbomb.* Stanford: Stanford University Press, 1976.

——. *Making Weapons, Talking Peace: A Physicist's Odyssey from Hiroshima to Geneva.* New York: Basic Books, 1987.

——. *Race to Oblivion.* New York: Simon & Schuster, 1970.

York, Herbert F., with Jerome Weisner. "National Security and the Nuclear Test Ban." *Scientific American* 211 (October 1964): 27–35.

## Books and Essays

Akin, William E. *Technocracy and the American Dream: The Technocratic Movement, 1900–1941.* Berkeley: University of California Press, 1977.

Alchon, Guy. *The Invisible Hand of Planning: Capitalism, Social Science, and the State in the 1920s.* Princeton: Princeton University Press, 1985.

Alperovitz, Gar. *Atomic Diplomacy: Hiroshima and Potsdam.* 1963. Reprint, London: Pluto Press, 1994.

Ambrose, Stephen. *Eisenhower, the President.* New York: Simon & Schuster, 1984.

Aron, Raymond. *The Century of Total War.* Boston: Beacon Press, 1954.

Atomic Energy Commission. *In the Matter of J. Robert Oppenheimer: Transcript of Hearing before Personnel Security Board and Texts of Principal Documents and Letters.* Cambridge, Mass.: MIT Press, 1971.

Balogh, Brian. *Chain Reaction: Expert Debate and Public Participation in American Commercial Nuclear Power, 1945–1975.* New York: Cambridge University Press, 1991.

——. "Reorganizing the Organizational Synthesis: Reconsidering Modern Federal-Professional Relations." *Studies in American Political Development* 5, no. 1 (1991): 119–72.

Baracca, Angelo. " 'Big Science' vs. 'Little Science': Laboratories and Leading Ideas in Conflict, Nuclear Physics in the Thirties and Forties." *Physis* 30, nos. 2–3 (1993): 373–90.

Beard, Edmund. *Developing the ICBM: A Study in Bureaucratic Politics.* New York: Columbia University Press, 1976.

Bender, Thomas. *Intellect and Public Life: Essays on the Social History of Academic Intellectuals in the United States.* Baltimore: Johns Hopkins University Press, 1993.

Bernstein, Barton. *The Atomic Bomb: The Critical Issues.* Boston: Little, Brown, 1976.

——. "Four Physicists and the Bomb: The Early Years." *Historical Studies in the Physical and Biological Sciences* 18, no. 2 (1988): 231–64.

——. "The Oppenheimer Case Reconsidered." *Stanford Law Review* 42 (July 1990): 1383–1484.

——. "Seizing the Contested Terrain of Early Nuclear History: Stimson, Conant, and Their Allies Explain the Decision to Use the Atomic Bomb." *Diplomatic History* 17 (Winter 1993): 35–72.

Bernstein, Jeremy. *Experiencing Science.* New York: Basic Books, 1978.

——. *Hans Bethe: Prophet of Energy.* New York: Basic Books, 1980.

Bledstein, Burton J. *The Culture of Professionalism: The Middle Class and the Development of Higher Education in America*. New York: Norton, 1976.

Blum, John Morton. *V Was for Victory: Politics and American Culture during World War II*. New York: Harcourt Brace Jovanovich, 1976.

Boyer, Paul. *By the Bomb's Early Light: American Thought and Culture at the Dawn of the Atomic Age*. New York: Pantheon Books, 1985.

Brandes, Stuart D. *American Welfare Capitalism, 1880–1940*. Chicago: University of Chicago Press, 1976.

Brinkley, Alan. *The End of Reform: New Deal Liberalism in Recession and War*. New York: Knopf, 1995.

———. "Writing the History of Contemporary America: Dilemmas and Challenges." *Daedalus* 113 (Summer 1984): 121–41.

Bromberg, Joan. *NASA and the Space Industry*. Baltimore: Johns Hopkins University Press, 1999.

Bruce, Robert V. *The Launching of Modern American Science*. Ithaca: Cornell University Press, 1987.

Bundy, McGeorge. *Danger and Survival: Choices about the Bomb in the First Fifty Years*. 1989. Reprint, New York: Vintage, 1990.

Calvert, Monte A. *The Mechanical Engineer in America, 1830–1910: Professional Cultures in Conflict*. Baltimore: Johns Hopkins Press, 1967.

Cantelon, Philip L., et al., eds. *The American Atom: A Documentary History of Nuclear Policies from the Discovery of Fission to the Present*. 2d ed. Philadelphia: University of Pennsylvania Press, 1991.

Carlisle, Rodney P., and Joan M. Zenzen. *Supplying the Nuclear Arsenal: American Production Reactors, 1942–1992*. Baltimore: Johns Hopkins University Press, 1996.

Carlson, W. Bernard. "Academic Entrepreneurship and Engineering Education: Dugald Jackson and the MIT-GE Cooperative Engineering Course, 1907–1932." *Technology and Culture* 29 (July 1988): 536–67.

Chandler, Alfred D., Jr. *The Visible Hand: The Managerial Revolution in American Business*. Cambridge, Mass.: Harvard University Press, 1977.

Childs, Herbert. *An American Genius: The Life of Ernest Orlando Lawrence*. New York: E. P. Dutton, 1968.

Clarfield, Gerald H., and William M. Wiecek. *Nuclear America: Military and Civilian Nuclear Power in the United States, 1940–1980*. New York: Harper & Row, 1984.

Clowse, Barbara Barksdale. *Brainpower for the Cold War: The Sputnik Crisis and the National Defense Education Act of 1958*. Westport, Conn.: Greenwood Press, 1981.

Cmiel, Kenneth. *Democratic Eloquence: The Fight over Popular Speech in Nineteenth-Century America*. Berkeley: University of California Press, 1990.

Cochran, Thomas C., and William Miller. *The Age of Enterprise: A Social History of Industrial America*. New York: Macmillan, 1942.

Cochrane, Rexmond. *The National Academy of Sciences: The First Hundred Years, 1863–1963*. Washington, D.C.: National Academy of Sciences, 1978.

Cohen, Lizabeth. *Making a New Deal: Industrial Workers in Chicago, 1919–1939*. New York: Cambridge University Press, 1990.

Cook, Blanche Wiesen. *The Declassified Eisenhower: A Divided Legacy of Peace and Political Warfare*. New York: Penguin, 1981.

Cowan, Ruth S. *More Work for Mother: The Ironies of Household Technology from the Open Hearth to the Microwave*. New York: Basic Books, 1983.

Critchlow, David T. *The Brookings Institution, 1916–1952: Expertise and the Public Interest in a Democratic Society*. DeKalb: Northern Illinois University Press, 1985.

Crunden, Robert. *Ministers of Reform: The Progressive Achievement in American Civilization, 1880–1920*. Urbana: University of Illinois Press, 1984.

Cuff, Robert D. "The Cooperative Impulse and War: The Origins of the Council of National Defense and Advisory Commission." In *Building the Organizational Society: Essays in Associational Activities in Modern America*, edited by Jerry Israel. New York: Free Press, 1972.

Danielian, Noobar R. *AT&T: The Story of Industrial Conquest*. 1939. Reprint, New York: Ayer, 1974.

Davis, Nuel Pharr. *Lawrence and Oppenheimer*. New York: Simon & Schuster, 1968.

Dawley, Alan. *Struggles for Justice: Social Responsibility and the Liberal State*. Cambridge, Mass.: Harvard University Press, 1991.

Degler, Carl. *In Search of Human Nature: The Decline and Revival of Darwinism in American Social Thought*. New York: Oxford University Press, 1991.

DeVorkin, David H. "Organizing for Space Research: The V-2 Rocket Panel." *Historical Studies in the Physical and Biological Sciences* 18, no. 1 (1987): 1–24.

———. *Science with a Vengeance: How the Military Created the U.S. Space Sciences after World War II*. New York: Springer-Verlag, 1992.

Dewey, John. *The Public and Its Problems*. New York: Henry Holt, 1927.

Divine, Robert A. *Blowing on the Wind: The Nuclear Test Ban Debate, 1954–60*. New York: Oxford University Press, 1978.

———. *The Sputnik Challenge*. New York: Oxford University Press, 1993.

Dockrill, Saki. *Eisenhower's New-Look National Security Policy, 1953–61*. New York: St. Martin's Press, 1996.

Douglas, Susan J. *Inventing American Broadcasting, 1899–1922*. Baltimore: Johns Hopkins University Press, 1987.

Dupree, A. Hunter. *Science and the Federal Government: A History of Policies and Activities to 1940*. Cambridge, Mass.: Harvard University Press, 1957.

Eakins, David W. "The Origins of Corporate Liberal Policy Research, 1916–1922: The Political-Economic Expert and the Decline of Public Debate." In *Building the Organizational Society: Essays in Associational Activities in Modern America*, edited by Jerry Israel. New York: Free Press, 1972.

Edwards, Paul N. *The Closed World: Computers and the Politics of Discourse in Cold War America*. Cambridge, Mass.: MIT Press, 1996.

England, J. Merton. *A Patron for Pure Science: The National Science Foundation's Formative Years, 1945–57*. Washington, D.C.: National Science Foundation, 1982.

Evangelista, Matthew. *Innovation and the Arms Race: How the United States and the Soviet Union Develop New Military Technologies*. Ithaca: Cornell University Press, 1988.

Fischer, Claude S. *America Calling: A Social History of the Telephone to 1940*. Berkeley: University of California Press, 1992.

Fleming, Donald, and Bernard Bailyn, eds. *The Intellectual Migration: Europe and America, 1930–1960*. Cambridge, Mass.: Harvard University Press, 1969.

Fones-Wolf, Elizabeth. *Selling Free Enterprise: The Business Assault on Labor and Liberalism, 1945–1960*. Urbana: University of Illinois Press, 1994.

Forman, Paul. "Behind Quantum Electronics: National Security as Basis for Physical Research in the United States, 1940–1960." *Historical Studies in the Physical and Biological Sciences* 18, no. 1 (1987): 149–229.

———. "Into Quantum Electronics: The Maser as 'Gadget' of Cold War America." In *National Military Establishments and the Advancement of Science and Technology*, edited by Paul Forman and Jose M. Sanchez-Ron. Boston Studies in the Philosophy of Science, 180. Dordrecht, The Netherlands: Kluwer Academic Publishers, 1996.

——. "Inventing the Maser in Postwar America." *Osiris*, 2d ser., 7 (1992): 105–34.

Fox, Richard Wrightman, and T. J. Jackson Lears, eds. *The Culture of Consumption: Critical Essays in American History, 1880–1980*. New York: Pantheon Books, 1983.

Franklin, H. Bruce. *War Stars: The Superweapon and the American Imagination*. New York: Oxford University Press, 1988.

Fredrickson, George M. *The Black Image in the White Mind: The Debate on Afro-American Character and Destiny, 1817–1914*. New York: Harper & Row, 1971.

Friedberg, Aaron L. *In the Shadow of the Garrison State: America's Anti-Statism and Its Cold War Grand Strategy*. Princeton: Princeton University Press, 2000.

Furner, Mary O. *Advocacy and Objectivity: A Crisis in the Professionalization of American Social Science, 1865–1905*. Lexington: University Press of Kentucky, 1975.

Gaddis, John Lewis. *The United States and the Origins of the Cold War, 1941–1947*. New York: Columbia University Press, 1972.

Galambos, Louis. "The Emerging Organizational Synthesis in Modern American History." *Business History Review* 44 (1970): 279–90.

——. *The Public Image of Big Business in America, 1880–1940: A Quantitative Study in Social Change*. Baltimore: Johns Hopkins University Press, 1975.

——. "Technology, Political Economy, and Professionalization: Central Themes of the Organizational Synthesis." *Business History Review* 57 (Winter 1983): 471–93.

——. "Theodore N. Vail and the Role of Innovation in the Modern Bell System." *Business History Review* 66 (Spring 1992): 95–126.

——, ed. *The New American State: Bureaucracies and Policies since World War II*. Baltimore: Johns Hopkins University Press, 1987.

Galambos, Louis, and Joseph Pratt. *The Rise of the Corporate Commonwealth*. New York: Basic Books, 1988.

Galison, Peter, and Barton J. Bernstein. "In Any Light: Scientists and the Decision to Build the Superbomb, 1952–1954." *Historical Studies in the Physical and Biological Sciences* 19, no. 2 (1989): 267–347.

Galison, Peter, and Bruce Hevly. *Big Science: The Growth of Large Scale Research*. Stanford: Stanford University Press, 1992.

Gardner, Lloyd. *Pay Any Price: Lyndon Johnson and the Wars for Vietnam*. Chicago: Ivan R. Dee, 1995.

Geiger, Roger. *Research and Relevant Knowledge: American Research Universities since World War II*. New York: Oxford University Press, 1993.

——. "Science, Universities, and National Defense, 1945–1970." *Osiris*, 2d ser., 7 (1992): 26–48.

——. *To Advance Knowledge: The Growth of American Research Universities, 1900–1940*. New York: Oxford University Press, 1986.

Gilbert, James. *Designing the Industrial State: The Intellectual Pursuit of Collectivism in America, 1880–1940*. Chicago: Quadrangle Press, 1972.

Gilpin, Robert. *American Scientists and Nuclear Weapons Policy*. Princeton: Princeton University Press, 1962.

Goldberg, Stanley. "Inventing a Climate of Opinion: Vannevar Bush and the Decision to Build the Bomb." *Isis* 83 (1992): 429–52.

Goldberg, Stanley, and Roger Stuewer. *The Michelson Era in American Science, 1870–1930*. New York: American Institute of Physics, 1988.

Goodchild, Peter. *J. Robert Oppenheimer, Shatterer of Worlds*. New York: Houghton Mifflin, 1980.

Goodman, Paul. *Like a Conquered Province*. New York: Random House, 1967.

——. *New Reformation: Notes of a Neolithic Conservative*. New York: Random House, 1970.

Goodstein, Judith R. *Millikan's School: A History of the California Institute of Technology.* New York: Norton, 1991.

Graebner, Norman A., ed. *The National Security: Its Theory and Practice, 1945–1960.* New York: Oxford University Press, 1986.

Greenberg, Daniel. *The Politics of Pure Science.* New York: New American Library, 1967.

Griffith, Robert. "Dwight D. Eisenhower and the Corporate Commonwealth." *American Historical Review* 87 (April 1982): 87–122.

———. "Forging America's Post-War Order: Domestic Politics and Political Economy in the Age of Truman." In *The Truman Presidency*, edited by Michael Lacey. New York: Cambridge University Press, 1989.

Groves, Leslie. *Now It Can Be Told.* New York: Harper & Row, 1962.

Gruber, Carol S. *Mars and Minerva: World War I and the Uses of the Higher Learning in America.* Baton Rouge: Louisiana State University Press, 1975.

———. "The Overhead System in Government Sponsored Academic Science: Origins and Early Development." *Historical Studies in the Physical and Biological Sciences* 25, no. 2 (1995): 241–68.

Haber, Samuel. *Authority and Honor in the American Professions, 1750–1900.* Chicago: University of Chicago Press, 1991.

———. *Efficiency and Uplift: Scientific Management in the Progressive Era, 1890–1920.* Chicago: University of Chicago Press, 1964.

Hamby, Alonzo. *Beyond the New Deal: Harry S. Truman and American Liberalism.* New York: Columbia University Press, 1973.

Harris, Neil. *Cultural Excursions: Marketing Appetites and Cultural Tastes in Modern America.* Chicago: University of Chicago Press, 1990.

———. "Cultural Institutions and American Modernization." *Journal of Library History* 16 (Winter 1981): 28–47.

———. "The Gilded Age Revisited: Boston and the Museum Movement." *American Quarterly* 14, no. 4 (1962): 545–66.

———. *Humbug: The Art of P. T. Barnum.* Chicago: University of Chicago Press, 1972.

———. "Museums, Merchandising and Popular Taste: The Struggle for Influence." In *Material Culture and the Study of American Life*, edited by Ian M. G. Quimby. New York: Norton, 1978.

Hart, David M. *Forged Consensus: Science, Technology, and Economic Policy in the United States, 1921–1953.* Princeton: Princeton University Press, 1998.

Haskell, Thomas. *The Emergence of Professional Social Science: The American Social Science Association and the Nineteenth Century Crisis of Authority.* Urbana: University of Illinois Press, 1977.

———, ed. *The Authority of Experts: Studies in History and Theory.* Bloomington: Indiana University Press, 1984.

Hawley, Ellis W. *The Great War and the Search for a Modern Order.* New York: St. Martin's Press, 1979.

———. "Herbert Hoover, the Commerce Secretariat and the Vision of an 'Associative State,' 1921–1928." *Journal of American History* 61 (1974): 116–40.

———. "Herbert Hoover and American Corporatism, 1929–1933." In *The Hoover Presidency: A Reappraisal*, edited by Martin Fausold and George T. Mazuzan. Albany: State University of New York Press, 1974.

Heilbron, J. L., and Robert Seidel. *Lawrence and His Laboratory.* Vol. 1: *A History of the Lawrence Berkeley Laboratory.* Berkeley: University of California Press, 1989.

Herken, Gregg. *Cardinal Choices: Presidential Science Advising from the Atomic Bomb to SDI.* New York: Oxford University Press, 1992.

———. *Counsels of War.* New York: Knopf, 1984.

———. *The Winning Weapon: The Atomic Bomb in the Cold War, 1945–1950*. 1980. Reprint, Princeton: Princeton University Press, 1988.

Hersey, John. *Hiroshima*. 1946. Reprint, New York: Vintage, 1989.

Hershberg, James. *James B. Conant: From Harvard to Hiroshima*. New York: Knopf, 1993.

———. " 'Over My Dead Body': James B. Conant and the Hydrogen Bomb." In *Science, Technology and the Military*, Vol. 1, edited by Everett Mendelsohn, Merritt Roe Smith, and Peter Weingart. Dordrecht, The Netherlands: Kluwer Academic Publishers, 1988.

Hewlett, Richard G., and Oscar E. Anderson Jr. *The New World: A History of the United States Atomic Energy Commission*. Vol. 1: *1939–1946*. 1962. Reprint, Berkeley: University of California Press, 1990.

Hewlett, Richard G., and Francis Duncan. *Atomic Shield: A History of the United States Atomic Energy Commission*. Vol. 2: *1947–1952*. 1969. Reprint, Berkeley: University of California Press, 1990.

Hewlett, Richard G., and Jack Holl. *Atoms for Peace and War: A History of the Atomic Energy Commission*. Vol. 3: *1953–1961*. Berkeley: University of California Press, 1989.

Hoch, Paul K. "The Crystallization of a Strategic Alliance: The American Physics Elite and the Military in the 1940s." In *Science, Technology and the Military*, Vol. 1, edited by Everett Mendelsohn, Merritt Roe Smith, and Peter Weingart. Dordrecht, The Netherlands: Kluwer Academic Publishers, 1988.

Hofstadter, Richard. *The Age of Reform: From Bryan to F.D.R.* New York: Random House, 1956.

———. *Anti-Intellectualism in American Life*. New York: Knopf, 1963.

———. *Social Darwinism in American Thought*. Boston: Beacon Press, 1955.

Hogan, Michael J. *A Cross of Iron: Harry S. Truman and the Origins of the National Security State, 1945–1954*. New York: Cambridge University Press, 1998.

Hollinger, David. *In the American Province: Studies in the History and Historiography of Ideas*. Baltimore: Johns Hopkins University Press, 1985.

———. *Science, Jews, and Secular Culture: Studies in Mid-Twentieth Century American Intellectual History*. Princeton: Princeton University Press, 1996.

Holloway, David. *Stalin and the Bomb: The Soviet Union and Atomic Energy, 1939–1956*. New Haven: Yale University Press, 1994.

Hooks, Gregory. *Forging the Military-Industrial Complex: World War II's Battle of the Potomac*. Urbana: University of Illinois Press, 1991.

Hounshell, David. "Du Pont and the Management of Large Scale Research and Development." In *Big Science: The Growth of Large Scale Research*, edited by Peter Galison and Bruce Hevly. Stanford: Stanford University Press, 1992.

Hounshell, David, and John Smith Jr. *Science and Corporate Strategy: Du Pont R&D, 1902–1980*. New York: Cambridge University Press, 1988.

Hughes, Thomas P. *American Genesis: A Century of Invention and Technological Enthusiasm, 1870–1970*. New York: Penguin, 1989.

Israel, Jerry, ed. *Building the Organizational Society: Essays in Associational Activities in Modern America*. New York: Free Press, 1972.

Israel, Paul. *From Machine Shop to Industrial Laboratory: Telegraphy and the Changing Context of American Invention, 1830–1920*. Baltimore: Johns Hopkins University Press, 1992.

Johnston, Marjorie, ed. *The Cosmos of Arthur Holly Compton*. New York: Knopf, 1967.

Jones, James H. *Bad Blood: The Tuskegee Syphilis Experiment*. New York: Free Press, 1981.

Jordan, John. *Machine-Age Ideology: Social Engineering and American Liberalism, 1911–1939*. Chapel Hill: University of North Carolina Press, 1994.

——. " 'Society Improved the Way You Can Improve a Dynamo': Charles Steinmetz and the Politics of Efficiency." *Technology and Culture* 30 (1989): 57–82.

Kamen, Martin D. *Radiant Science, Dark Politics*. Berkeley: University of California Press, 1985.

Kargon, Robert H., ed. *The Maturing of American Science*. Washington, D.C.: American Association for the Advancement of Science, 1974.

Kargon, Robert H., and Elizabeth Hodes. "Karl Compton, Isaiah Bowman, and the Politics of Science in the Great Depression." *Isis* 76 (1985): 301–18.

Karl, Barry. "Presidential Planning and Social Science Research: Mr. Hoover's Experts." *Perspectives in American History* 3 (1969): 347–409.

——. *The Uneasy State: The U.S. from 1915–1945*. Chicago: University of Chicago Press, 1983.

Karsten, Peter. "Armed Progressives: The Military Reorganizes for the American Century." In *Building the Organizational Society: Essays in Associational Activities in Modern America*, edited by Jerry Israel. New York: Free Press, 1972.

Kasson, John. *Civilizing the Machine*. New York: Penguin, 1976.

Katz, Milton S. *Ban the Bomb: A History of SANE, the Committee for a Sane Nuclear Policy*. New York: Praeger, 1986.

Keller, Morton. *Affairs of State: Public Life in Late Nineteenth Century America*. Cambridge, Mass.: Harvard University Press, 1977.

——. *Regulating a New Economy: Public Policy and Economic Change in America, 1900–1930*. Cambridge, Mass.: Harvard University Press, 1990.

Kennan, George F. *Memoirs*. Vol. 1: *1925–1950*. Boston: Little, Brown, 1967.

——. *Memoirs*. Vol. 2: *1950–1963*. Boston: Little, Brown, 1972.

Kennedy, David. *Over Here: The First World War and American Society*. New York: Oxford University Press, 1980.

Kevles, Daniel. "Cold War and Hot Physics: Science, Security, and the American State, 1945–56." *Historical Studies in the Physical and Biological Sciences* 20, no. 2 (1989): 239–64.

——. *In the Name of Eugenics: Genetics and the Uses of Human Heredity*. Berkeley: University of California Press, 1986.

——. "K1S2: Korea, Science and the State." In *Big Science: The Growth of Large Scale Research*, edited by Peter Galison and Bruce Hevly. Stanford: Stanford University Press, 1992.

——. "The National Science Foundation and the Debate over Postwar Research Policy, 1942–1945." *Isis* 68 (March 1977): 5–26.

——. *The Physicists: A History of a Professional Community*. 1976. Reprint, Cambridge, Mass.: Harvard University Press, 1987.

——. "R&D and the Arms Race: An Analytical Look." In *Science, Technology and the Military*, Vol. 1, edited by Everett Mendelsohn, Merritt Roe Smith, and Peter Weingart. Dordrecht, The Netherlands: Kluwer Academic Publishers, 1988.

——. "Scientists, the Military, and the Control of Postwar Defense Research: The Case of the Research Board for National Security, 1944–1946." *Technology and Culture* 16 (January 1975): 20–47.

Killian, James R. *Sputnik, Scientists, and Eisenhower*. Cambridge, Mass.: MIT Press, 1977.

Kistiakowsky, George. *A Scientist at the White House*. Cambridge, Mass.: Harvard University Press, 1976.

Kleinman, Daniel Lee. *Politics on the Endless Frontier: Postwar Research Policy in the United States*. Durham: Duke University Press, 1995.

Kline, Ronald. "Construing 'Technology' as 'Applied Science': Public Rhetoric of Scientists and Engineers in the United States, 1880–1945." *Isis* 86 (June 1995): 194–221.

———. *Steinmetz: Engineer and Socialist*. Baltimore: Johns Hopkins University Press, 1992.

Kloppenberg, James T. *Uncertain Victory: Social Democracy and Progressivism in European and American Thought, 1870–1920*. New York: Oxford University Press, 1986.

Kohler, Robert E. *Partners in Science: Foundations and Natural Scientists, 1900–1945*. Chicago: University of Chicago Press, 1991.

Kolko, Gabriel. *The Triumph of Conservatism*. New York: Free Press, 1963.

Koppes, Clayton R. *JPL and the American Space Program*. New Haven: Yale University Press, 1982.

Kuhn, Thomas. *The Structure of Scientific Revolutions*. Chicago: University of Chicago Press, 1970.

Kuznick, Peter J. *Beyond the Laboratory: Scientists as Political Activists in 1930s America*. Chicago: University of Chicago Press, 1987.

Lacey, Michael, ed. *The Truman Presidency*. New York: Cambridge University Press, 1989.

LaFollette, Marcel C. *Making Science Our Own: Public Images of Science, 1910–1955*. Chicago: University of Chicago Press, 1991.

Laurence, William L. *Dawn over Zero*. New York: Knopf, 1946.

Layton, Edwin. *The Revolt of the Engineers*. 1971. Reprint, Baltimore: Johns Hopkins University Press, 1986.

Leach, William. *Land of Desire: Merchants, Power and the Rise of a New American Culture*. New York: Vintage, 1993.

Lears, T. J. Jackson. "The Concept of Cultural Hegemony." *American Historical Review* 90 (June 1985): 567–93.

———. *Fables of Abundance: A Cultural History of Advertising in America*. New York: Basic Books, 1994.

———. *No Place of Grace: Anti-Modernism and the Transformation of American Culture, 1880–1920*. New York: Pantheon Books, 1981.

Lecuyer, Christophe. "The Making of a Science-Based Technological University: Karl Compton, James Killian, and the Reform of MIT, 1930–1957." *Historical Studies in the Physical and Biological Sciences* 23, no. 1 (1992): 153–80.

———. "MIT, Progressive Reform, and 'Industrial Service,' 1890–1920." *Historical Studies in the Physical and Biological Sciences* 23, no. 2 (1993): 279–300.

Leffler, Melvyn P. *A Preponderance of Power: National Security, the Truman Administration, and the Cold War*. Stanford: Stanford University Press, 1992.

Leslie, Stuart W. *The Cold War and American Science: The Military-Industrial-Academic Complex at MIT and Stanford*. New York: Columbia University Press, 1993.

———. "Playing the Education Game to Win: The Military and Interdisciplinary Research at Stanford." *Historical Studies in the Physical and Biological Sciences* 18, no. 1 (1987): 55–88.

———. "Profit and Loss: The Military and MIT in the Postwar Era." *Historical Studies in the Physical and Biological Sciences* 21, no. 1 (1990): 59–85.

Levine, David O. *The American College and the Culture of Aspiration, 1915–1940*. Ithaca: Cornell University Press, 1986.

Lifton, Robert Jay, and Eric Markusen. *The Genocidal Mentality: Nazi Holocaust and Nuclear Threat*. New York: Basic Books, 1990.

Lippmann, Walter. *Drift and Mastery*. 1914. Reprint, Madison: University of Wisconsin Press, 1985.

———. *The Phantom Public*. New York: Harcourt Brace, 1925.

Livingston, James. *Pragmatism and the Political Economy of Cultural Revolution, 1850–1940*. Chapel Hill: University of North Carolina Press, 1994.

Lowen, Rebecca S. *Creating the Cold War University: The Transformation of Stanford*. Berkeley: University of California Press, 1997.

———. "Transforming the University: Administrators, Physicists, and Industrial and Federal Patronage at Stanford, 1935–1949." *History of Education Quarterly* 31, no. 3 (1991): 365–88.

Lustig, R. Jeffrey. *Corporate Liberalism: The Origins of Modern American Political Theory, 1890–1920*. Berkeley: University of California Press, 1982.

McCartin, Joseph A. *Labor's Great War: The Struggle for Industrial Democracy and the Origins of Modern American Labor Relations, 1912–1921*. Chapel Hill: University of North Carolina Press, 1997.

McCraw, Thomas K. *TVA and the Power Fight, 1933–1939*. Philadelphia: J. B. Lippincott, 1971.

McCurdy, Howard E. *Inside NASA: High Technology and Organizational Change in the U.S. Space Program*. Baltimore: Johns Hopkins University Press, 1993.

McDougall, Walter. *The Heavens and the Earth: A Political History of the Space Age*. New York: Basic Books, 1985.

McMahon, A. Michal. *The Making of a Profession: A Century of Electrical Engineering in America*. New York: IEEE Press, 1984.

McQuaid, Kim. "Competition, Cartellization and the Corporate Ethic: General Electric's Leadership during the New Deal Era, 1933–1940." *American Journal of Economics and Sociology* 36, no. 3 (1977): 323–34.

———. "Corporate Liberalism in the American Business Community, 1920–1940." *Business History Review* 52 (Autumn 1978): 342–68.

———. *Uneasy Partners: Big Business in American Politics, 1945–1990*. Baltimore: Johns Hopkins University Press, 1994.

Marchand, Roland. *Creating the Corporate Soul: The Rise of Public Relations and Corporate Imagery in American Big Business*. Berkeley: University of California Press, 1998.

Marchand, Roland, and M. L. Smith. "Corporate Science on Display." In *Scientific Authority and Twentieth-Century America*, edited by Ronald G. Walters. Baltimore: Johns Hopkins University Press, 1997.

Marx, Leo. *The Machine in the Garden: Technology and the Pastoral Ideal in America*. New York: Oxford University Press, 1964.

———. *The Pilot and the Passenger: Essays on Literature, Technology and Culture in the United States*. New York: Oxford University Press, 1988.

May, Elaine Tyler. *Homeward Bound: American Families in the Cold War Era*. New York: Basic Books, 1988.

May, Henry F. *The End of American Innocence: The First Years of Our Own Time, 1912–1917*. New York: Oxford University Press, 1959.

Millard, Andre. *Edison and the Business of Innovation*. Baltimore: Johns Hopkins University Press, 1990.

Millis, Walter, with Harvey C. Mansfield and Harold Stein. *Arms and the State: Civil-Military Elements in National Policy*. New York: Twentieth Century Fund, 1958.

Mills, C. Wright. *The Power Elite*. New York: Oxford University Press, 1956.

Montgomery, David. *The Fall of the House of Labor: The Workplace, the State, and American Labor Activism, 1865–1925*. New York: Cambridge University Press, 1987.

Mukerji, Chandra. *A Fragile Power: Scientists and the State*. Princeton: Princeton University Press, 1989.

Mumford, Lewis. *The Myth of the Machine*. Vol. 2: *The Pentagon of Power*. New York: Harcourt Brace Jovanovich, 1970.

———. *Technics and Civilization*. 1934. Reprint, New York: Harcourt Brace Jovanovich, 1963.

National Academy of Sciences. *National Academy of Sciences of the United States of America, Biographical Memoirs*. Vol. 18. Washington, D.C.: National Academy of Sciences, 1938.

Needell, Allan A. "From Military Research to Big Science: Lloyd Berkner and Science-Statesmanship in the Postwar Era." In *Big Science: The Growth of Large Scale Research*, edited by Peter Galison and Bruce Hevly. Stanford: Stanford University Press, 1992.

———. "Lloyd Berkner, Merle Tuve, and the Federal Role in Radio Astronomy." *Osiris* 3 (1987): 261–88.

———. "Preparing for the Space Age: University-Based Research, 1946–1957." *Historical Studies in the Physical and Biological Sciences* 18, no. 1 (1987): 89–109.

———. "Rabi, Berkner, and the Rehabilitation of Science in Europe: The Cold War Context of American Support for International Science, 1945–1958." In *The United States and the Integration of Europe: Legacies of the Postwar Era*, edited by Francis H. Heller and John Gillingham. New York: St. Martin's Press, 1996.

Nelson, Daniel. *Frederick W. Taylor and the Rise of Scientific Management*. Madison: University of Wisconsin Press, 1980.

Neu, Charles E. "The Rise of the National Security Bureaucracy." In *The New American State: Bureaucracies and Policies since World War II*, edited by Louis Galambos. Baltimore: Johns Hopkins University Press, 1987.

Neufeld, Jacob. *Ballistic Missiles in the United States Air Force, 1945–1960*. Washington, D.C.: Office of Air Force History, United States Air Force, 1990.

Neuse, Steven M. *David E. Lilienthal: The Journey of an American Liberal*. Knoxville: University of Tennessee Press, 1996.

Newman, James R., and Byron S. Miller. *The Control of Atomic Energy*. New York: Whittlesey House, 1948.

Noble, David F. *America by Design: Science, Technology and the Rise of Corporate Capitalism*. New York: Oxford University Press, 1977.

———. *Forces of Production: A Social History of Automation*. New York: Oxford University Press, 1984.

Numbers, Ronald L., and Charles E. Rosenberg. *The Scientific Enterprise in America: Readings from Isis*. Chicago: University of Chicago Press, 1996.

Nye, David E. *American Technological Sublime*. Cambridge, Mass.: MIT Press, 1994.

———. *Consuming Power: A Social History of American Energies*. Cambridge, Mass.: MIT Press, 1998.

———. *Electrifying America: Social Meanings of a New Technology*. Cambridge, Mass.: MIT Press, 1990.

———. *Image Worlds: Corporate Identities at General Electric, 1890–1930*. Cambridge, Mass.: MIT Press, 1985.

Ohmann, Richard. *Selling Culture: Magazines, Markets, and Class at the Turn of the Century*. London: Verso, 1996.

Oppenheimer, J. Robert. *The Open Mind*. New York: Simon & Schuster, 1955.

Owens, Larry. "The Counterproductive Management of Science in the Second World War: Vannevar Bush and the Office of Scientific Research and Development." *Business History Review* 68 (Winter 1994): 515–76.

——. "MIT and the Federal 'Angel': Academic R&D and Federal-Private Cooperation before World War II." *Isis* 81 (1990): 189–213.

——. "Vannevar Bush and the Differential Analyzer: The Text and Context of an Early Computer." *Technology and Culture* 27 (1986): 63–95.

Page, Arthur. *The Bell Telephone System*. New York: AT&T, 1941.

Page, Arthur, et al. *Modern Communication*. New York: Houghton Mifflin, 1932.

Pfau, Richard. *No Sacrifice Too Great: The Life of Lewis L. Strauss*. Charlottesville: University Press of Virginia, 1984.

Platt, Harold L. *The Electrical City: Energy and the Growth of the Chicago Area, 1880–1930*. Chicago: University of Chicago Press, 1991.

Price, Don K. *Government and Science*. New York: New York University Press, 1954.

Price, Matt. "Roots of Dissent: The Chicago Met Lab and the Origins of the Franck Report." *Isis* 86 (1995): 222–44.

Pursell, Carroll. "Science Agencies in World War II: The OSRD and Its Challengers." In *The Sciences in the American Context: New Perspectives*, edited by Nathan Reingold. Washington, D.C.: Smithsonian Institution Press, 1979.

Quimby, Ian M. G., ed. *Material Culture and the Study of American Life*. New York: Norton, 1978.

Reich, Leonard S. "Irving Langmuir and the Pursuit of Science and Technology in the Corporate Environment." *Technology and Culture* 24 (April 1983): 199–221.

——. *The Making of American Industrial Research: Science and Business at GE and Bell, 1876–1926*. New York: Cambridge University Press, 1985.

Reingold, Nathan. "Choosing the Future: The U.S. Research Community, 1944–1946." *Historical Studies in the Physical and Biological Sciences* 25, no. 2 (1995): 301–28.

——. *Science American Style*. New Brunswick: Rutgers University Press, 1991.

——. "Vannevar Bush's New Deal for Research: Or, the Triumph of the Old Order." *Historical Studies in the Physical and Biological Sciences* 17, no. 2 (1987): 299–344.

——, ed. *The Sciences in the American Context: New Perspectives*. Washington, D.C.: Smithsonian Institution Press, 1979.

Reynolds, Terry S., ed. *The Engineer in America: A Historical Anthology from Technology and Culture*. Chicago: University of Chicago Press, 1991.

Rhodes, F. L. *John J. Carty: An Appreciation*. New York: Privately printed, 1932.

Rhodes, Richard. *Dark Sun: The Making of the Hydrogen Bomb*. New York: Simon & Schuster, 1995.

——. *The Making of the Atomic Bomb*. New York: Simon & Schuster, 1986.

Rigden, John S. *Rabi, Scientist and Citizen*. New York: Basic Books, 1987.

Rodgers, Daniel T. *Atlantic Crossings: Social Politics in a Progressive Age*. Cambridge, Mass.: Harvard University Press, 1998.

——. *The Work Ethic in Industrial America, 1850–1920*. Chicago: University of Chicago Press, 1978.

Roland, Alex. *Model Research: A History of the National Advisory Committee for Aeronautics, 1915–1958*. Washington, D.C.: NASA, 1985.

Roman, Peter J. *Eisenhower and the Missile Gap*. Ithaca: Cornell University Press, 1995.

Rosenberg, Charles. "Science and American Social Thought." In *Science and Society in the United States*, edited by David Van Tassel and Michael Hall. Homewood, Ill.: Dorsey Press, 1966.

Rosenberg, David Alan. "The Origins of Overkill: Nuclear Weapons and American Strategy." In *The National Security: Its Theory and Practice, 1945–1960*, edited by Norman A. Graebner. New York: Oxford University Press, 1986.

Ross, Dorothy. "A Historian's View of American Social Science." In *Scientific Authority and Twentieth-Century America*, edited by Ronald G. Walters. Baltimore: Johns Hopkins University Press, 1997.

——. "Modernist Social Science in the Land of the New/Old." In *Modernist Impulses in the Human Sciences, 1870–1930*, edited by Dorothy Ross. Baltimore: Johns Hopkins University Press, 1994.

——. *The Origins of American Social Science*. New York: Cambridge University Press, 1991.

——, ed. *Modernist Impulses in the Human Sciences, 1870–1930*. Baltimore: Johns Hopkins University Press, 1994.

Ross, Malcolm, ed., with Maurice Holland and William Spragen. *Profitable Practice in Industrial Research: Tested Principles of Research Laboratory Organization, Administration and Operation*. New York: Harper & Brothers, 1932.

Roszak, Theodore. *The Making of a Counterculture: Reflections on the Technocratic Society and Its Youthful Opposition*. Garden City, N.Y.: Anchor Books, 1969.

Russert, Cynthia Eagle. *Darwin in America: The Intellectual Response, 1865–1912*. San Francisco: W. H. Freeman, 1976.

Rydell, Robert. " 'The Fan Dance of Science': America's World's Fairs in the Great Depression." *Isis* 76, no. 283 (1985): 525–42.

——. *World of Fairs: The Century-of-Progress Expositions*. Chicago: University of Chicago Press, 1993.

Sanders, Jerry W. *Peddlers of Crisis: The Committee on the Present Danger and the Politics of Containment*. Boston: South End Press, 1983.

Sapolsky, Harvey. *Science and the Navy: The History of the Office of Naval Research*. Princeton: Princeton University Press, 1990.

Schatz, Ronald. *The Electrical Worker: A History of Labor at General Electric and Westinghouse, 1923–1960*. Urbana: University of Illinois Press, 1983.

Schell, Jonathan. *The Abolition*. New York: Knopf, 1984.

——. *The Fate of the Earth*. New York: Knopf, 1982.

Schlesinger, Arthur M., Jr. *The Vital Center: The Politics of Freedom*. New York: Da Capo, 1988.

Schrecker, Ellen W. *No Ivory Tower: McCarthyism and the Universities*. New York: Oxford University Press, 1986.

Schweber, Silvan S. "The Empiricist Temper Regnant: Theoretical Physics in the United States, 1920–1950." *Historical Studies in the Physical and Biological Sciences* 17, no. 1 (1986): 55–98.

——. "The Mutual Embrace of Science and the Military: ONR and the Growth of Physics in the United States after World War II." In *Science, Technology and the Military*, Vol. 1, edited by Everett Mendelsohn, Merritt Roe Smith, and Peter Weingart. Dordrecht, The Netherlands: Kluwer Academic Publishers, 1988.

Segal, Howard P. *Technological Utopianism in American Culture*. Chicago: University of Chicago Press, 1985.

Seidel, Robert. "Accelerating Science: The Postwar Transformation of the Lawrence Radiation Laboratory." *Historical Studies in the Physical and Biological Sciences* 13, no. 2 (1983): 375–400.

——. "From Glow to Flow: A History of Military Laser Research and Development." *Historical Studies in the Physical and Biological Sciences* 18, no. 1 (1987): 111–47.

——. "A Home for Big Science: The Atomic Energy Commission's Laboratory System." *Historical Studies in the Physical and Biological Sciences* 16, no. 1 (1986): 135–75.

Servos, John. "The Industrial Relations of Science: Chemistry at MIT, 1900–1939." *Isis* 71, no. 259 (1980): 531–49.

——. *Physical Chemistry from Ostwald to Pauling: The Making of a Science in America.* Princeton: Princeton University Press, 1990.

Sherry, Michael. *In the Shadow of War: The United States since the 1930s.* New Haven: Yale University Press, 1995.

——. *Preparing for the Next War: American Plans for Postwar Defense, 1941–1945.* New Haven: Yale University Press, 1977.

——. *The Rise of American Air Power: The Creation of Armageddon.* New Haven: Yale University Press, 1987.

Sherwin, Martin J. *A World Destroyed: The Atomic Bomb and the Grand Alliance.* New York: Vintage, 1977.

Sklar, Martin J. *The Corporate Reconstruction of American Capitalism, 1890–1916.* New York: Cambridge University Press, 1988.

Skowronek, Stephen. *Building a New American State: The Expansion of National Administrative Capacities, 1877–1920.* New York: Cambridge University Press, 1982.

Smith, Alice Kimball. *A Peril and a Hope: The Scientists' Movement in America, 1945–47.* Chicago: University of Chicago Press, 1965.

Smith, Greg David. *The Anatomy of a Business Strategy: Bell, Western Electric, and the Origins of the American Telephone Industry.* Baltimore: Johns Hopkins University Press, 1985.

Smith, John Kenley, Jr. "The Scientific Tradition in American Research." *Technology and Culture* 31 (January 1990): 121–31.

Smith, Michael L. "Selling the Moon: The U.S. Manned Space Program and the Triumph of Commodity Scientism." In *The Culture of Consumption: Critical Essays in American History, 1880–1980,* edited by Richard Wrightman Fox and T. J. Jackson Lears. New York: Pantheon Books, 1983.

Smyth, Henry DeWolf. *Atomic Energy for Military Purposes.* 1945. Reprint, Stanford: Stanford University Press, 1989.

Steel, Ronald. *Walter Lippmann and the American Century.* New York, Vintage, 1981.

Steidle, Barbara C. " 'Reasonable' Reform: The Attitude of Bar and Bench toward Liability Law and Workmen's Compensation." In *Building the Organizational Society: Essays in Associational Activities in Modern America,* edited by Jerry Israel. New York: Free Press, 1972.

Stern, Philip, with Harold Green. *The Oppenheimer Case: Security on Trial.* New York: Harper & Row, 1969.

Sylves, Richard T. *Nuclear Oracles: A Political History of the General Advisory Committee of the Atomic Energy Commission, 1947–1977.* Ames: Iowa State University Press, 1987.

Talbert, Roy. *FDR's Utopian: Arthur Morgan of the TVA.* Jackson: University Press of Mississippi, 1987.

Teitelbaum, Robert. *Profits of Science: The American Marriage of Business and Technology.* New York: Basic Books, 1994.

Tichi, Cecelia. *Shifting Gears: Technology, Literature, Culture in Modernist America.* Chapel Hill: University of North Carolina Press, 1987.

Tobey, Ronald C. *The American Ideology of National Science, 1919–1930.* Pittsburgh: University of Pittsburgh Press, 1971.

——. *Technology as Freedom: The New Deal and the Electrical Modernization of the American Home.* Berkeley: University of California Press, 1996.

Tucker, William H. *The Science and Politics of Racial Research.* Urbana: University of Illinois Press, 1994.

Van Tassell, David, and Michael Hall, eds. *Science and Society in the United States.* Homewood, Ill.: Dorsey Press, 1966.

Veysey, Lawrence. *The Emergence of the American University*. Chicago: University of Chicago Press, 1965.

Walker, Mark. *Nazi Science: Myth, Truth, and the German Atomic Bomb*. New York: Plenum, 1995.

Walters, Ronald G., ed. *Scientific Authority and Twentieth-Century America*. Baltimore: Johns Hopkins University Press, 1997.

Wang, Jessica. *American Science in an Age of Anxiety: Scientists, Anti-Communism, and the Cold War*. Chapel Hill: University of North Carolina Press, 1999.

———. "Liberals, the Progressive Left, and the Political Economy of Post-War American Science: The National Science Foundation Debate Revisited." *Historical Studies in the Physical and Biological Sciences* 26, no. 1 (1995): 139–66.

Warner, Sam Bass. *Province of Reason*. Cambridge, Mass.: Harvard University Press, 1984.

Weart, Spencer R. *Nuclear Fear: A History of Images*. Cambridge, Mass.: Harvard University Press, 1985.

Weaver, Warren. *Scenes of Change: A Lifetime in American Science*. New York: Scribners, 1970.

Webb, James E. *Space Age Management: The Large-Scale Approach*. New York: McGraw-Hill, 1969.

Weinstein, James B. *The Corporate Ideal in the Liberal State, 1900–1918*. Boston: Beacon Press, 1968.

Weisskopf, Victor. *The Joy of Insight: Passions of a Physicist*. New York: Basic Books, 1991.

White, Morton. *The Revolt against Formalism*. New York: Oxford University Press, 1949.

Whitfield, Stephen J. *The Culture of the Cold War*. 2d ed. Baltimore: Johns Hopkins University Press, 1996.

Wiebe, Robert. *Businessmen and Reform: A Study of the Progressive Movement*. Chicago: Elephant Paperback, 1989.

———. *The Search for Order, 1877–1920*. New York: Hill & Wang, 1967.

———. *Self-Rule: A Cultural History of American Democracy*. Chicago: University of Chicago Press, 1995.

Wilson, Jane. *All in Our Time: The Reminiscences of Twelve Nuclear Pioneers*. Chicago: Bulletin of Atomic Scientists, 1975.

Wilson, Joan Hoff. *Herbert Hoover: Forgotten Progressive*. Boston: Little, Brown, 1975.

Winkler, Allan M. *Life under a Cloud: American Anxiety about the Atom*. New York: Oxford University Press, 1993.

Wise, George. "A New Role for Professional Scientists in Industry: Industrial Research at General Electric, 1900–1916." *Technology and Culture* 21 (July 1980): 408–29.

———. *Willis R. Whitney, GE, and the Origins of American Industrial Research*. New York: Columbia University Press, 1985.

Wittner, Lawrence S. *The Struggle against the Bomb*. 2 vols. Stanford: Stanford University Press, 1993.

Wyden, Peter. *Day One*. New York: Simon & Schuster, 1984.

Yergin, Daniel. *Shattered Peace: The Origins of the Cold War and the National Security State*. Boston: Houghton Mifflin, 1977.

Zachary, G. Pascal. *Endless Frontier: Vannevar Bush, Engineer of the American Century*. New York: Free Press, 1997.

Zeiger, Robert H. "Herbert Hoover, the Wage-Earner, and the 'New Economic System,' 1919–1929." *Business History Review* 61 (Summer 1977): 161–89.

Zunz, Olivier. *Making America Corporate, 1870–1920*. Chicago: University of Chicago Press, 1990.

——. "Producers, Brokers, and Users of Knowledge: The Institutional Matrix." In *Modernist Impulses in the Human Sciences, 1870–1930*, edited by Dorothy Ross. Baltimore: Johns Hopkins University Press, 1994.